Water resources professionals have an obligation to conceive and manage water resource systems such that they will fully contribute to an improved quality of life for all humans, now and into the future. Those water resource systems that will be able to satisfy the changing demands that will inevitably be placed on them, without significant system degradation, can be called 'sustainable'.

This volume examines the practical issues and challenges raised by the concept of sustainability as it is applied to water resource system design and mangement. An international group of experts have reviewed various guidelines for achieving greater degrees of sustainability and the extent to which they have been applied in a number of case studies. Approaches for measuring and modelling sustainability are provided. Ways in which these measures and models might be used when evaluating alternative designs and operating policies are illustrated.

This monograph will be particularly valuable for practising engineers and planners, and as a supplementary text for graduate students in civil and environmental engineering, hydrology, geography and economics.

DANIEL P. LOUCKS is a professor in the Department of Civil and Environmental Engineering at Cornell University.

JOHN S. GLADWELL is president of Hydro Tech International in Vancouver.

The **International Hydrological Programme** (IHP) was established by the United Nations Educational, Scientific and Cultural Organisation (UNESCO) in 1975 as the successor to the International Hydrological Decade. The long-term goal of the IHP is to advance our understanding of processes occurring in the water cycle and to integrate this knowledge into water resources management. The IHP is the only UN science and educational programme in the field of water resources, and one of its outputs has been a steady stream of technical and information documents aimed at water specialists and decision-makers.

The **International Hydrology Series** has been developed by the IHP in collaboration with Cambridge University Press as a major collection of research monographs, synthesis volumes and graduate texts on the subject of water. Authoritative and international in scope, the various books within the Series all contribute to the aims of the IHP in improving scientific and technical knowledge of fresh water processes, in providing research know-how in stimulating the responsible management of water resources.

EDITORIAL ADVISORY BOARD
Secretary to the Advisory Board:
Dr Michael Bonell   *Division of Water Sciences, UNESCO, 1 rue Miollis, Paris 75732, France*

Members of the Advisory Board:
Professor B.P.F. Braga Jr.   *Cento Technológica de Hidráulica, Sao Paulo, Brazil*
Professor G. Dagan   *Faculty of Engineering, Tel Aviv University, Israel*
Dr J. Khouri   *Water Resources Division, Arab Centre for Studies of Arid Zones and Dry Lands, Damascus, Syria*
Dr G. Leavesley   *U.S. Geological Survey, Water Resources Division, Denver Federal Center, Colorado, USA*
Dr E. Morris   *British Antarctic Survey, Cambridge, United Kingdom*
Professor L. Oyebande   *Department of Geography and Planning, University of Lagos, Nigeria*
Professor S. Sorooshian   *College of Engineering and Mines, University of Arizona, Tucson, USA*
Professor K. Takeuchi   *Department of Civil and Environmental Engineering, Yamanashi University, Japan*
Professor D.E. Walling   *Department of Geography, University of Exeter, United Kingdom*
Dr I. White   *CSIRO Division of Environmental Mechanics, Canberra, Australia*

TITLES IN PRINT IN THE SERIES
M. Bonell, M.M. Hufschmidt and J.S. Gladwell   *Hydrology and Water Management in the Humid Tropics: Hydrological Research Issues and Strategies for Water Management*
Z.W. Kundzewicz   *New Uncertainty Concepts in Hydrology*
R.A. Feddes   *Space and Time Scale Variability and Interdependencies in the Various Hydrological Processes*
J. Gibert, J. Mathieu and F. Fournier   *Groundwater and Surface Water Ecotones: Biological and Hydrological Interactions and Management Options*
G. Dagan and S. Neuman   *Subsurface Flow and Transport: A Stochastic Approach*
D.P. Loucks and J.S. Gladwell   *Sustainability Criteria for Water Resource Systems*

INTERNATIONAL HYDROLOGY SERIES

Sustainability Criteria for Water Resource Systems

Daniel P. Loucks, chair and John S. Gladwell, editor

CAMBRIDGE
UNIVERSITY PRESS

333·91

T2159)

PUBLISHED BY THE PRESS SYNDICATE OF THE UNIVERSITY OF CAMBRIDGE
The Pitt Building, Trumpington Street, Cambridge, United Kingdom

CAMBRIDGE UNIVERSITY PRESS
The Edinburgh Building, Cambridge CB2 2RU, UK http://www.cup.cam.ac.uk
40 West 20th Street, New York, NY 10011-4211, USA http://www.cup.org
10 Stamford Road, Oakleigh, Melbourne 3166, Australia

First published 1999

Printed in the United Kingdom at the University Press, Cambridge

Typeset in 9.5/13pt Times by Keyword Typesetting Services Ltd, Wallington, Surrey

A catalogue record for this book is available from the British Library

Library of Congress Cataloguing in Publication data

Sustainability criteria for water resource systems / prepared by the
   working group of UNESCO/IHP Project M-4.3, Daniel P Loucks,
   chairman ... [et al.] ; John S. Gladwell, editor.
        p.   cm. – (International hydrology series)
      UNESCO, International Hydrological Programme.
      "March 1997."
      Includes bibliographical references (p.).
      ISBN 0 521 56044 6 (hardbound)
      1. Water resources development–Systems engineering–Case studies.
   2. Water-supply engineering–Case studies.   3. Sustainable
   development–Case studies.   I. Loucks, Daniel P.   II. Gladwell,
   John S.   III. Unesco/IHP-IV Project M-4.3.   IV. International
   Hydrological Programme.   V. Series
   TC409.S8   1999
   333.91'15–dc21   98-29507 CIP

ISBN 0 521 56044 6 hardback

Contents

Preface   ix
Acknowledgments   xiii
A brief overview   1

1   Introduction   6

2   Sustainability issues and challenges   9

3   Defining sustainability   26

4   Measuring sustainability   32

5   Sustainability guidelines and case studies   42

6   Economic sustainability criteria   67

7   Ecological and environmental sustainability
    criteria   81

8   Institutional and social aspects of sustainability   91

9   Sustainability and modeling technology   98

10 Sustainability, hydrologic risk and uncertainty   108

11 Equity, education and technology transfer   122

12 Conclusion   131

    References   134
    Index   138

Preface

Water resource professionals have an obligation to design and manage water resource systems so that they can fully contribute to an improved quality of life for all humans. Water resource systems that are able to satisfy the changing demands placed on them, now and on into the future, without system degradation, can be called 'sustainable'.

This document examines the issues and challenges raised by the concept of sustainability applied to water resource system design and management. It reviews some of the guidelines that have been suggested for achieving a greater degree of sustainability and the extent to which they have been applied in a number of case studies. It outlines some approaches for measuring and modeling sustainability and illustrates ways in which these measures and models might be used when evaluating alternative designs and operating policies.

This monograph was prepared by a working group of the International Hydrological Programme of the United Nations Scientific, Educational and Cultural Organization (UNESCO) with contributions from the Task Committee of the Division of Water Resources Planning and Management of the American Society of Civil Engineers (ASCE). Both groups were formed to explore ways in which the concept of sustainability might be used as a measure of system performance when evaluating alternative water resource plans and management policies.

This monograph is for those interested and involved in the planning and management of water resources. It was written by a group of individuals sharing a background in environmental and water resources systems planning and management, but having differing experiences and opinions. Hence different points of view are presented with the hope that they will stimulate thinking about just how water resource sys-

tems should be developed and managed, not only for those living today but also for those who will be dependent on these systems in the future.

Sustainability is a unifying concept that emphasizes the need to consider the long-term future as well as the present. This includes the future economic, environmental, ecological, physical and social impacts that will result from decisions and actions taken today. While we cannot know with certainty what all these impacts will be, or what future generations of individuals or societies will want or value, we can attempt to predict what we think might happen and what future generations may want or value as we develop our current plans, designs and management policies. Admittedly we can only guess at what future generations would like us to do now in our generation for them in their generations. We must take these guesses into account as we make our decisions or take actions to satisfy our immediate demands and desires.

Because sustainability is a function of various economic, environmental, ecological, social and physical goals and objectives, analyses must inevitably involve multi-objective tradeoffs in a multi-disciplinary and multi-participatory decision-making process. No single discipline, and certainly no single profession or interest group, has the wisdom to make these tradeoffs. They can only be determined through a political process involving all interested and impacted stakeholders. The participants must at least attempt to take into account the likely preferences of those not able to be present in this decision-making process, namely those who will be living in the future and who will be impacted by current resource management decisions.

Sustainability is intimately related to various measures of risk and uncertainly about a future we cannot know but which we can surely influence. Clearly our guesses about the future will, with certainty, be wrong. Hence they will need to be revised periodically. Recognizing that some management objectives will change over time, we must consider the adaptability or robustness of the systems we design and operate today to this management uncertainty and to the changes in the quantity and quality of the resource being managed.

We begin the monograph with a discussion of the definition of sustainability that is commonly used, but which we think is not very helpful for water resources planning and management. Nevertheless, this common definition is often cited, used and discussed in the considerable literature that exists today on the subject of sustainability and sustainable development. In Chapter 2 we review some of the major issues and challenges posed by this commonly accepted definition of sustainability and try to identify why sustainability has been so difficult to quantify and to define very precisely. The discussion of the challenges and issues associated with this broad concept of sustainability (as applied to various water resources purposes) leads us to the particular definition we propose for water resource systems planning and management.

In Chapter 3 we define sustainability in a way that seems appropriate to those of us involved in water resources planning and management. This definition allows us, in Chapter 4, to identify and examine in more detail some ways of measuring sustainability for selected water resources functions or purposes. These measures rely on the inputs and judgments of those having an interest in such systems. The interests of different stakeholders may differ. While we recognize computer analyses lie behind most decisions involving facility design, construction and operation, it is their development and use toward achieving a shared common vision among all stakeholders that is important with respect to system sustainability.

One result of this growing concern over the need to achieve system sustainability (however defined) has been the creation of a number of guidelines for its achievement. The guidelines, summarized in Chapter 5, were created by various professional engineering organizations in various countries. The primary purpose of these guidelines is to help those in the engineering profession, and especially the practicing engineer, design and manage systems that are sustainable. Systems can be designed and managed so as to be sustainable even though particular projects and even institutions making up those systems may not be. There are many aspects of sustainable systems, including change, but all should lead toward the development and use of more sus-

tainable technologies, to more sustainable environments and ecosystems, to more sustainable economic and financial policies, to more sustainable institutions and societies, and to improved long-term human health and welfare.

Chapter 5 also includes a series of brief descriptions – case studies – of some water resource development and management projects. They serve to illustrate the extent to which sustainability criteria have or have not been achieved in particular situations. These real-world examples also illustrate the difficult tradeoffs that must be made among various goals and interest groups when designing water resource systems and implementing policies for managing them.

Chapter 6 explores some economic criteria and associated models that consider economic objectives. The discussion examines the issue of appropriate discount rates – the weight we put on our current assets compared to the weights we assign to assets of future generations. One of the major needs still unmet is our ability to value non-monetary goods (such as those derived from our environment and our ecosystems) in monetary terms. Since we have not yet learned how this can be done, we are forced to make comparisons and tradeoffs between economic and environmental or ecological criteria expressed in different metrics (units of measure). Chapter 7 identifies some of these environmental and ecological criteria and associated modeling and implementation approaches aimed at achieving a greater degree of sustainability. Through a series of examples emphasis is placed on the importance of communication and public participation, hopefully leading to a shared vision in what the outcome should be. The adaptive approach to planning is shown to be extremely useful, with guidelines given for its implementation.

Individuals and societies manage themselves through their institutions. While this monograph is not focused on that aspect of sustainability in any great detail, Chapter 8 briefly reviews some of the social and institutional aspects of sustainability, illustrating their importance for water resources planners and managers. It is, after all, through the institutions of our world that decisions are made regarding our water resources development and use. They are the *clients* for all of our planning and design of structures, and of all our modeling and decision support systems development. Institutions can foster and encourage increasing sustainability, or they can do just the opposite. Three examples are given. Chapter 9 examines how a variety of modeling technologies can and should contribute toward higher levels of sustainability. Emphasis is given to ways of improving the planning and management process and the information upon which recommendations are made and decisions are based.

The concept of Decision Support Systems (DSSs) is presented and emphasized.

Chapter 10 examines how economic, environmental and hydrologic risk and uncertainty impacts on our attempts to define and work with sustainability criteria. Anyone involved in water resources planning and management must contend with risk and uncertainty. No one can look into the future with precision. Both the future supplies of water and the future demands for the services provided by water resource systems are unknown at the time system design and operating decisions are made. Professionals are asked to provide for this uncertain future. As a result, system robustness and other risk-based measures of system performance become important considerations, and are intimately tied into any measure of sustainability.

Chapter 11 addresses some equity, education and technology transfer issues related to sustainability. While the discussionw > is brief, the subject is as important as any discussed in this monograph. Different individuals will have different views as to what is equitable or ethical. The correct view is not always obvious.

Particularly for those in educational institutions, it is important to consider capacity building and the technology transfer issues with respect to sustainability. Some of the education, training and technology transfer aspects of sustainability are examined. Chapter 11 also discusses the important roles professional societies as well as those of educational institutions in producing and providing the expertise needed to continue into the future the efforts being made today toward achieving more sustainable systems.

The monograph concludes with Chapter 12 that highlights some key points concerning the planning and management of sustainable water resource systems. It emphasizes that in our search for sustainable development, the effectiveness of any mechanism derived to reach that goal depends, in the end, on the quality of the individuals interested in pursuing it.

Acknowledgments

We are grateful to many individuals for various inputs. We especially want to thank Graeme Dandy of the University of Adelaide and Gary Codner of Monash University in Australia for their suggestions and contributions. Wil Thissen of the Technical University in Delft and his students and colleagues together with Rob Klomp and Jos Dijkman of the Delft Hydraulic Laboratory in the Netherlands were particularly helpful in identifying some important issues and sources of information on this subject. Professors Juan Valdes and Ralph Wurbs at Texas A&M University together with J. Marco at the University of Valencia and A. Mejia at the World Bank contributed directly to the chapter on Risk and Sustainability. We have also drawn heavily from the COWAR report of Jordaan *et al.* (1993). For all their contributions and for the support we obtained from UNESCO, from the US Committee on Scientific Hydrology of the United States Geological Survey (the US 'IHP committee'), from the American Society of Civil Engineers (ASCE), and from the German National Committee for UNESCO who sponsored our Karlsruhe Symposium in June, 1994, we are most thankful.

The UNESCO working group consisted of the following members:

A. Andreu, Universidad Politecnica de Valencia, Spain

J.J. Bogardi, UNESCO, Paris, France
J.S. Gladwell, Hydro Tech International, Vancouver, Canada
Y.Y. Haimes, University of Virginia, USA
S. Kaden, WASY, GmbH, Germany
D.P. Loucks, Cornell University, USA (Chairman)
J. Kindler, World Bank, USA
H.-P. Nachtnebel, Universität für Bodenkultur, Austria
E. Plate, University of Karlsruhe, Germany
S.P. Simonovic, University of Manitoba, Canada
U. Shamir, Technion, Israel
E. Todini, University of Bologna, Italy

The ASCE Task Committee on Sustainability Criteria included:

M.J. Bender, University of Manitoba, Canada
C.D.D. Howard, Charles Howard & Assoc. Ltd., Canada
D.P. Loucks, Cornell University, USA (Chairman)
W.A. Lyon, University of Pennsylvania, USA
G.F. McMahon, Camp Dresser & McKee, Inc., USA
K.E. Schilling, Institute for Water Resources, US Army Corps of Engineers, USA
J. Scott, Northern Colorado Water Conservancy District, USA
W. Viessman, University of Florida, USA

A brief overview

DEFINITION AND SCOPE

Sustainable water resource systems are those designed and managed to meet the needs of people living in the future as well as those of us living today. It is a philosophical concept. It is not a precise state of being. Sustainability criteria force us to consider the long-term future as well as the present. The actions that we as a society take now to satisfy its own needs and desires should not only depend on what those actions will do for us but on how they will affect our descendants as well. This consideration of the long-term impacts of current actions on future generations is the essence of sustainable development.

The concept of 'sustainability' can mean different things to different people. It seems to defy a precise definition. While the debate over just what sustainability means will continue, and questions over just what it is that should be sustained may remain unanswered, this need not delay our attempts to work towards the creation of more sustainable water resource systems – systems that can better serve our future generations while at the same time meeting our current goals and demands.

The concept of environmental and ecological sustainability has largely resulted from a growing concern about the long-run health of our planet. There is increasing evidence that our present resource use and management activities and actions, even at local levels, can significantly affect the welfare of those living within much larger regions in the future. Water resource management problems can no longer be justly viewed as purely technical and of interest only to those living within the individual watersheds where those problems exist. Rather they must be seen as being closely related to broader societal structures, demands and issues. Many local water resources development and management projects may need to be viewed in a much more multi-disciplinary and inter-regional perspective than previously done.

Water resource management professionals spend a lot of time planning, designing and managing water resource systems to best meet the needs and objectives of those depending on or benefiting from these systems. Various criteria are used to compare and evaluate alternative plans specifying what, when, where and how much to do, and why. Sustainability indicators should be included among these evaluation criteria. Sustainability criteria will force us to assess the various impacts of our proposed plans, policies and practices on future generations as well as on our own generation.

What would future generations like us to do for them? What would they suggest we do today that will benefit them in the future? We don't know, but we can guess. We can only guess at the objectives or desires of future generations, and at what they would like us to do, now, for them living in this world sometime in the future. As uncertain as these guesses will be, we should make them and then take them into account as we act to satisfy our own immediate needs, demands and desires.

There may be tradeoffs between what we wish to do for ourselves today, or in our current generation, and what we think future generations might wish us to do for them. These tradeoffs, if any, between what present and future generations would like must be first identified. It is our job as professionals to do this. Once identified, or at least estimated, just what tradeoffs should be made can be debated and decided in the political arena. There is no scientific theory to help us identify which tradeoffs, if any, are optimum.

SUSTAINABILITY AND CHANGE

The inclusion of sustainability criteria along with the more common economic, environmental, ecological and social criteria used to evaluate alternative water resource development and management strategies may identify a need to change how we commonly develop and use our water resources. We need to consider change itself. Change over time is certain, just what it will be is the only thing that is uncertain. These changes will impact the physical, biological and social dimensions of water resource systems. An essential aspect in the planning, design and management of sustainable systems is the anticipation of change: changes in the natural system due to geomorphologic processes, in the engineered components due to aging, in the demands or desires due to a changing society, and even in the supply of water, possibly due to a changing climate. Change is an essential feature of sustainable water resources development and management. The notion that certain engineering structures, or systems of structures, and even social institutions, must continue to exist in the future as they are today is not a necessary condition for sustainability. Sustainable water resource systems are those designed and operated in ways that make them more adaptive, robust, and resilient to these uncertain changes. Sustainable water resource systems must be capable of effectively functioning under conditions of changing supplies, management objectives, and demands. Sustainable systems, like any others, may fail, but when they fail they must be capable of recovering and operating properly without undue costs.

In the face of certain changes, but with uncertain impacts, an evolving and adaptive strategy for water resources development, management and use is a necessary condition of sustainable development. Conversely, inflexibility in the face of new information and new objectives and new social and political environments is an indication of reduced system sustainability. Adaptive management is a process of adjusting management actions and directions, as appropriate, in light of new information on the current and likely future condition of our total environment and on our progress toward meeting our goals and objectives. Management decisions can be viewed as experiments, subject to modification – but with goals clearly in mind. Adaptive management recognizes the limitations of current knowledge and experience and that we learn by experimenting. It helps us move toward meeting our changing goals over time in the face of this incomplete knowledge and uncertainty. It accepts the fact that there is a continual need to review and revise environmental and other restoration and management approaches because of the changing as well as uncertain nature of our socioeconomic and natural environments.

Changes in the social and institutional components of water resource systems are often the most challenging because they involve changing the way individuals think and act. As individuals change, so may their institutions – the rules under which society functions. Sustainability requires that public and private institutions change over time in ways that are responsive to the changing demands of individuals.

Understanding how institutions are structured and function can help one understand better how water resource system development policies and operating rules might be altered when they become deficient, and who has the authority to change such rules, and in what ways. But to understand fully the boundaries of relevant institutions, water resource professionals must understand how institutions function under stress or under pressures for and against change from individuals within and outside the institution.

To be sustainable, a project must perform reliably during processes of change. The transition to new technologies, new management practices, and new institutions (or institutional leadership) must proceed in an orderly and equitable manner. Continuity and confidence in the new systems are prerequisites for sustainability, as are a proper respect for operation rules and for maintenance of the physical infrastructure.

SUSTAINABILITY AND SCALE

If we maintain too broad an interpretation of sustainable development it becomes difficult to determine progress toward achieving it. In particular, concern only with the sustainability of larger regions could overlook the unique attributes of particular local economies, environments, ecosystems, resource substitution and human health. On the other hand, not every hectare of land or every reach of every stream need be sustainable or self-sufficient. Not every watershed need be sustainable or self-sufficient. Even at river-basin or regional levels, it may not be possible to meet the 'needs' or demands of even the current generation let alone future generations if those needs or demands are greater than can be obtained on a continuing basis at acceptable economic, environmental and social costs. This highlights the need to consider the appropriate spatial scales when applying sustainability criteria to specific water resource systems.

We also need to consider the appropriate temporal scales when applying sustainability criteria to specific water resource systems. The achievement of higher levels of water resource system sustainability does not imply there will never be periods of time in the future in which the level of welfare derived from those systems decreases. Given the variations in natural water supplies – the fact that floods and droughts do occur – it is impossible, or at least very costly, to design and operate water resource systems that will never 'fail'. During periods of 'failure' the economic benefits derived from such systems may decrease. The ecological benefits may in fact depend on these events. One of the challenges of developing sustainability criteria is to identify the appropriate temporal scales over which progress in obtaining higher levels of welfare should be measured.

SUSTAINABILITY INDICES AND GUIDELINES

Sustainability indices provide ways we can measure relative levels of sustainability. They can be defined in a number of ways. One way is to express relative levels of sustainability as separate or weighted combinations of reliability, resilience and vulnerability measures of various criteria that contribute to human welfare. These criteria can be economic, environmental, ecological and social. To do this one must first identify the overall set of criteria and then for each one decide which ranges of values are satisfactory and which ranges are not. These decisions are subjective. They are generally based on human judgment or social goals, not scientific theory. In some cases they may be based on well-defined health standards, for example, but most criteria will not have predefined or published standards or threshold values separating what is considered satisfactory and what is not. For many criteria, the time duration as well as the extent of individual and cumulative failure may be important.

Guidelines for the development and management of sustainable water resource systems can be defined with respect to (1) the design, management and operation of physical infrastructure; (2) the quality of the environment or the health of ecosystems; (3) economics and finance; (4) institutions and society; (5) human health and welfare; and (6) planning and technology. The general public as well as professionals need to participate in the process of defining these guidelines. All concerned stakeholders should be involved in decision making. Bottom up approaches involving interested stakeholders who reap the benefits and, for the most part, pay the costs, are usually prerequisites for sustained system development, management and maintenance.

The identification and evaluation of sustainable development options requires a consideration of multiple criteria, not all of which can be expressed in monetary terms. What criteria would future generations use, and what weights would they place on them? We do not know, but it would seem reasonable to at least guess at what our descendants would want us to do for them. What kind of water resource systems would future generations like to inherit and manage as they work towards an even better quality of life for themselves? Of course this will depend on their objectives or goals. Since we do not know what these will be, we can assume that they may be the same as ours. Ignorance about the future is not an excuse for inaction or myopic decision-making. But since we cannot look into the future with precision, and since our planning methods that consider sustainability require us to make guesses about the future, our planning and decision-making process should be sequential and adaptive. The assumptions concerning the future need to be re-evaluated each time planning or decision-making takes place. Sustainability is a relative state. Achieving higher levels requires continual monitoring, adaptation and decision making.

One should not equate sustainability with the preservation of non-renewable resources. The central question in these cases is just how much of a non-renewable resource should be used now and on into the future, recognizing that while the future is important, it is also very uncertain. The trick is to use non-renewable resources, such as fossil groundwaters, during times when it is most beneficial. These time periods are not easy to determine in an uncertain future.

Important guidelines for the planning and management of sustainable water resource systems include:

- Developing a shared vision of desired social, economic and environmental goals benefiting present as well as future generations and identifying ways in which all parties can contribute to achieving that shared vision.
- Developing coordinated approaches among all concerned and interested agencies to accomplish these goals, collaborating with all stakeholders in recognition of mutual concerns.
- Using approaches that restore or maintain economic vitality, environmental quality and natural ecosystem biodiversity and health.
- Supporting actions that incorporate sustained economic, socio-cultural and community goals.
- Respecting and ensuring private property rights while meeting community goals and working cooperatively

with private stakeholders to accomplish these common and shared goals.

- Recognizing that economies, ecosystems and institutions are complex, dynamic (changing) and typically heterogeneous over space and time and developing management approaches that take into account and adapt to these characteristics.
- Integrating the best science available into the decision-making process, while continuing scientific research to improve knowledge and understanding.
- Establishing baseline conditions for system functioning and sustainability against which change can be measured.
- Monitoring and evaluating actions to determine if goals and objectives are being achieved.

SUSTAINABILITY AND TECHNOLOGY ————

All stakeholders involved in or impacted by the planning and management of water resources can be aided by the use of modern information processing technology. This technology includes computer-based interactive optimization and simulation models and programs, all specifically developed to perform more comprehensive multi-sector, multi-purpose, multi-objective water resources planning and management studies. Without such models, programs and associated data bases it would be extremely difficult to predict the expected future impacts of any proposed plan and management policy. Without the development and use of decision support systems incorporating these models, programs and data bases within an interactive menu-driven, graphics-based framework, it would be difficult for many to use these tools and data bases to explore their individual ideas, to test various assumptions, and to understand the output of their analyses.

Whatever modeling technology is developed and implemented to study a particular water resource system, it cannot address sustainability issues unless it addresses or simulates the variables of concern to those who will be affected by its management. Thus, the value of a particular predictive technology lies not only in its economic viability and its technical soundness but also in its adaptation to the local institutional and cultural environment. Decision support systems (DSSs) can provide all stakeholders concerned with a means to examine and study the various impacts, at various levels of detail, associated with any proposed decision regarding the use and management of water resource systems. DSSs can also support an adaptive, real-time planning and manage-

ment approach, in which the decisions as well as the DSSs themselves, can be updated and improved over time.

SUSTAINABILITY AND RISK ————

Sustainability implies a condition in which the frequency and severity of threats to society are decreasing over time. It implies a condition in which our environment and ecosystems are being managed in a way that prepares people to cope with stresses when they occur. Variability in water flows and qualities is a natural phenomena and must be preserved if such systems are to sustain their natural, or near natural, ecosystems. However, very extreme events typically bring substantial economic damages. Thus, the prevention, management and control of very extreme events have a high priority in the achievement of sustainability. Yet it is usually neither politically feasible nor economically possible to eliminate all potential hazards or to design all water resources systems to withstand any conceivable extreme event. Of interest, then, is the effectiveness of recovering from such events.

In risk assessment the analyst should be attempting to answer the following: What could go wrong? What is the likelihood that it will go wrong? What would be the consequences? The analyst then builds on the assessment process by seeking answers to a second set of questions: What can be done, i.e., what options are available for hazard reduction and response? What are the associated tradeoffs in terms of all costs, benefits and risks? What are the impacts of current management decisions on future options? Sustainability criteria must include risk measures and management as part of the overall assessment of possible system failures and their possible consequences. Water resource systems risk assessment and risk management planning should involve all having an interest in or who are impacted by those systems.

Long-term demand management involving land use and conservation programs can promote the efficient use of water continuously during normal as well as extreme conditions such as floods and droughts. The effect of drought on public water supplies necessitates cooperation between water users and local, regional and national public officials. But, since droughts are infrequent in many areas, water managers are faced with dealing with situations for which they typically have little or no past experience. Developing a national or regional drought policy and plan, then, is essential for reducing societal vulnerability and hence increasing system sustainability. Flood management and planning must not only take into account the risks of potential economic and social (psychological) damages resulting from flooding, but also the ecological and economical benefits of alternative flood plain development and use, and how it can be done to reduce potential damages.

SUSTAINABILITY AND TRAINING ————————

A key to the sustainable development of water resources is the existence of sufficiently well-trained personnel in all of the disciplines needed in the planning and development processes. In regions or localities where such a capacity is needed but does not exist, it should be developed. Training and education is a key input, and requirement, of sustainable development. While outside experts and aid organizations can provide temporary assistance, each major river basin region must inevitably come to depend primarily on its own professionals to provide the know-how and experience required for water resources development and management.

ACHIEVING SUSTAINABILITY ————————

Everyone involved in water resource systems development has an obligation to see that those systems provide sufficient quantities and qualities, at acceptable prices and reliabilities, and at the same time protect the environment and preserve the biodiversity and health of ecosystems for future generations. If our current water resources development and management practices result in degraded environments and ecosystems, those particular water resource systems will surely not be sustainable. There are many examples today of where this has happened. Would these 'failures' have occurred if sustainability criteria were considered when decisions were made? Are those who develop and manage water resource systems to meet today's expressed demands for food and fiber and economic livelihood considering the impact of their actions on future generations and their expected demands?

Any motivation to consider the future depends on the ability and willingness to understand the interactions of processes on very different spatial and temporal scales. It also depends on an informed and supportive public. Those who are managing natural resources need to ensure that the public as well as their representatives who make decisions are aware of the short and long-term temporal as well as spatial impacts and tradeoffs.

Given the uncertainty of what future generations will want, and the economic, environmental and ecological problems they will face, a guiding principle for the achievement of sustainable water resource systems is to maintain the options available to future generations. What we do now should interfere as little as possible with the proper functioning of natural life cycles within the watershed. Throughout the water resource system planning and management process, it is necessary to identify and include within the set of evaluation criteria all the beneficial and adverse ecological, economic, environmental and social effects – especially the long-term effects – associated with any proposed project.

Whatever is done to increase the level of sustainability of our water resources infrastructure will likely involve some costs or require some reduction in the immediate benefits those of us living today could receive. For example, if those living now had to pay for the required remedial measures of any contamination that they produce, they would be less likely to produce it. It is clear that wherever possible the prevention of pollution in excess of the receiving systems assimilative capacity is preferable to, and cheaper than, the reduction or elimination of its consequences. The challenge is to create the incentives that result in pollution prevention – that result in behavior that leads to higher levels of sustainability.

Water resources development and management is typically a public sector activity. Yet the money that is needed to develop and manage water, sustainably or otherwise, generally comes from the private sector. The money needed to create jobs, lift people out of poverty, and provide for the demands of growing populations comes from economic growth, domestic saving and wise investments at the national and international levels. While private profit-motivated businesses cannot be expected to achieve sustainable systems, economies and environments by themselves, an increasing number of them see that it is in their long-term interests to be partners with governments and non-governmental organizations, as appropriate, working together toward achieving this goal.

Everyone makes water management and use decisions, not just the professionals and the politicians. However, it is the job of the professionals, i.e., individuals who would be reading a book like this, to provide the information upon which informed decisions can be made. As our knowledge increases and as conditions and expectations change, so will our decisions. Professionals, particularly engineers, can contribute to sustainable development in two ways: by introducing environmentally beneficial practices within their own organizations, and by insuring that their projects not only meet their client's needs but at the same time contribute positively to sustainable development.

1 Introduction

SUSTAINABILITY: A UNIFYING HOLISTIC FOCUS

When the history of natural resource management during the last quarter of the 20th century is written, *sustainability* may well prove to be the major unifying concept that was advanced, discussed, promoted and accepted, even at the highest levels of government throughout most of the world. This is rather amazing, given the fact that as yet there is no consensus on its precise meaning or even how to measure it. The Brundtland Commission's report *Our Common Future* (WCED, 1987) promotes the all-encompassing concept of *sustainable development*. To quote:

> Humanity has the ability to make development sustainable to ensure that it meets the needs of the present without compromising the ability of future generations to meet their own needs.

In fact, achieving sustainable development that *meets the needs of the present without compromising the ability of future generations to meet their own needs* may never be realized, or even adequately quantified, but it is clearly a goal worthy of serious consideration.

The concept of sustainable resource use has been around for some time. Watershed and groundwater managers had been taught the principles of sustained yield management long before such publications as *Limits to Growth* (Meadows *et al.*, 1974) or *Our Common Future* (WCED, 1987). Farmers, fishermen and foresters have concerned themselves for some time now with how to achieve sustainable yields of food, fish and fiber, respectively, from a given area or region. The current concept of sustainable development, however, is much broader in scope than the term *sustained resource use* or *yield management* (see, for example,

Biswas, Jellali & Stout, 1993; Brooks, 1992; Brown, 1991; Goodland, Daly & El Serafy, 1991; Pearce & Warford, 1993; Plate, 1992; and Svedin, 1988). Today sustainable development refers to a process in which the economy, environment and ecosystem of a region develop in harmony and in a way that will improve over time. It is a concept so broad that it seems to defy a precise quantitative definition – a definition that anyone can use to measure the relative degree to which some action or policy contributes to a sustainable improvement in social welfare.

Historians reporting on the events of this last quarter century may show that the real value of books like *Limits to Growth* and *Our Common Future*, and conferences like the UN Conference on the Human Environment held in Stockholm in 1972, and the UN Conference on Environment and Development (UNCED) held in Rio de Janeiro in 1992, was not that they led to specific actions that turned out to have saved our planet. Their value may well be that they changed the way many viewed the environment and ecosystems as they worked to advance economic development and equity. We may look back to this last quarter century as a time when many changed their way of making decisions and began doing things in a manner that recognized, and was more compatible with, environmental resource realities and limits.

Perhaps what has pushed the concept of sustainability into the public's conscience more than any other factor is a growing awareness of the global scale of many environmental impacts associated with our economic development activities. What people do to the rain-forests of Brazil or Cambodia can affect those living in China and North America, and indeed, the consumption of coal, oil and wood for energy in China and North America will influence the decisions of resource managers living in Brazil and Cambodia. While some debate is still taking place regarding the extent of any global impacts associated with specific economic development activities, we know, and can easily document, the extent of various detrimental environmental impacts over large regions. People in stressed regions are justifiably concerned, especially those who have few if any choices as they strive to support their families. All of us are together on this planet, and the planet is not getting any bigger. It is thus incumbent upon us to manage our resources more effectively to permit continued development in this finite world – development not necessarily of more material goods, but of a higher quality of life for everyone.

The Brundtland Commission argued that economic development and the maintenance of a high quality environment need not be in conflict, provided humanity ensures that such economic development is sustainable. As stated by Bruce (1992):

> First, development must not damage or destroy the basic life support system of our planet earth: the air, the water and the soil, and the biological systems. Second, development must be economically sustainable to provide a continuous flow of goods and services derived from the Earth's natural resources, and thirdly it requires sustainable social systems, at international, national, local and family levels, to ensure the equitable distribution of the benefits of the goods and services produced, and of sustained life supporting systems.

The Brundtland Commission identified a direction for development and a road map to an acceptable future. Their report and a series of preparatory meetings ultimately lead to the United Nations Conference on Environment and Development (UNCED) held in Rio de Janeiro in June 1992. Sustainability, as discussed in that conference, has since became a household word, at least among those addressing both development and environment at the national and international levels. The development and management of water resource systems is a key part of sustainable development. Sustainable development cannot succeed without sustainable water resource systems supporting that development. While the UNCED conference, and all the preparatory conferences and meetings leading up to the UNCED conference, identified some broad guidelines and principles for water resources development and management, they did not define or translate these broad guidelines into specific concepts that can be applied to the designing, operating and maintaining of water resources and water projects in specific regions. The details of how water resources systems should be developed and managed so

as to contribute to the fullest extent possible to sustainable economic and social development in specific regions can only be worked out by all interested and impacted stakeholders in those regions. Documents resulting from various national or international conferences and meetings, and even working groups or committees such as this monograph, can only provide some assistance and guidance to those who are actually involved in planning and decision-making in specific regions.

There now exists a well-stocked bag of tools available for studying and analyzing problems of supply and demand and for planning, designing and operating facilities that meet the demands for water of sufficient quantity and quality at reasonable costs. Research and the collective experience of generations of water engineers have provided professionals with many of the requisite methods. However, over time, conditions and objectives change. Well-established design concepts must be reviewed, revised and adapted to meet current and expected future conditions. New methods must also be developed and tested to meet new objectives and new demands of society. One of these new (or at least renewed) demands is that of achieving sustainability. Meeting sustainability objectives will certainly require an increased understanding of the interactions of nature and society. Even with greater understanding, however, a catalogue of problems arises for which answers must be found. Obtaining those answers will require investments in research and technological development.

A primary consequence of sustainability is that the single- or multiple-purpose project approach must be expanded to one that is more integrated or holistic, multi-disciplinary and regional in its view. The different purposes of water resource development – water supply, water quality, flood control, navigation, recreation, hydropower, etc. – should no longer be considered as merely technical or economic issues that can be examined locally (and separately) by professionals. Water resources problems must not be viewed as purely technical or economic single-purpose and single-project (or even multi-purpose, multi-project) challenges and opportunities. Rather, they must be seen as planning and management activities that are intimately intertwined with broader societal demands and issues. While engineering is needed, engineering itself is only one of many disciplines that must be involved in holistic, integrated and sustainable water resources development and management. Water resource systems must be considered integral parts of a changing societal system. The interactions of the system with society and environment must be taken into account by experts from all appropriate disciplines.

Systems must obviously be designed and managed to meet the current needs or demands of those who use or can benefit from the use of the system. They must also be sufficiently flexible (robust) so that, if required, they can easily adapt to unexpected future changes in demands or purposes. Systems must also be resilient, which implies an ability of structures or institutions, with proper maintenance, to recover and function properly after some unforeseen failure event. A resilient system, if damaged by a rare event, can be reconstructed with a minimum of effort, and at affordable costs.

As valuable and critical as water is for economic development and human health, it is commonly considered a free resource. Water is one of the most essential, and most subsidized, inputs to human activity throughout the world. In the more developed regions, people strive to live where they can find the work and quality of life they want. If the quantity and/or quality of water is insufficient, their public institutions are often expected to, and usually do, provide the needed quantity and quality of water at acceptable (subsidized) costs. From the financial point of view these systems may be unsustainable.

that will be physically, economically, environmentally, ecologically and socially acceptable and beneficial by current as well as future generations. How can this be done? What criteria can guide us toward achieving more sustainable water resources systems? This monograph addresses these questions. While it focuses on the contributions scientists, engineers, economists, planners and other specialists can make, it recognizes that their contributions are not sufficient by themselves. Important contributions to efficient and sustainable water management and use must also come from the public, the stakeholders and their political representatives and institutions.

Most of today's problems can be solved, and in a way that is sustainable. But leadership, together with an appropriate management structure, must exist before this can happen. If specialists in one or more areas of water resources planning and management are to become leaders in this transition to a more sustainable management of natural resources and environment, they must become involved in the political processes that take place in decision-making institutions.

WHY THIS MONOGRAPH?

Our capacity to move earth, build dams, pump, treat, distribute water, and produce hydropower, has increased over the past century. So has our capacity to manage and control water supplies to meet a multiplicity of purposes and objectives. But this increase in control has been accompanied by increases in economic and environmental costs. The costs of the increased control include those associated with the loss of natural habitats, the increase in threats to supplies from pollutants, the increasing frequency of demand-related droughts, and the mining of groundwater aquifers. As a result, it has become evident that many of our water resource developments and management practices implemented during this past century should be re-examined. We need to be sure that what has been and is now being done with our water and money will result in a sustainable (long-term) improvement in the quality of life for those who are dependent upon that water, now and on into the future.

This monograph addresses the need and challenge to re-examine our approaches to water resources planning and management. If the results are to be sustainable, we need to develop a more holistic and integrated life-cycle approach to water resources planning, development and management. Such an approach should lead to plans, facilities and policies

SOME THINGS TO REMEMBER

- The concept of sustainability has grown in importance because of the growing awareness of the global scale of the environmental impacts of economic development activities.
- There already exists a well-stocked bag of tools available for studying and analyzing problems of supply and demand and for planning, designing and operating facilities.
- The single or multiple project approach must be expanded to one that is more integrated or holistic, multi-disciplinary and regional in its view.
- Water resources problems must not be viewed as purely technical or economic single-purpose or even multi-project challenges and opportunities. Rather, they must be seen as being intimately intertwined with broader societal demands and issues.
- Systems must be designed to be robust and resilient so that they can easily adapt to unexpected future changes and be able to recover and function properly after some unforeseen failure event.
- Even though most of today's problems can be solved in a way that is sustainable, a proper management structure must exist before this can happen.

2 Sustainability issues and challenges

THE BROAD VIEW

Most descriptions of sustainable water resources development include three broad considerations or viewpoints:

- *Nature* – the view that river–aquifer–estuarine systems and their environment and ecosystems have value in their own right and hence efforts to protect these resources, habitats and biodiversity are warranted.
- *Current generation* – the recognition that those who live today have demands they expect to be met from water resource systems. These demands and goals differ over space. While addressing these varied demands and goals there must also be a move toward reducing the gap in the quality of life that may exist among peoples within a river basin or watershed or even among multiple basins or watersheds. The management of water resources, it is believed, can be directed towards increasing the degree of equity and justice.
- *Future generations* – the view that our descendants have the right to at least the same if not a better economy, environment, and quality of life than we enjoy today. Current actions with regard to the development and management of water resource systems should not eliminate options future generations may wish to take to fulfill their demands and goals.

The word 'sustainability' implies the continuance or maintenance of a certain situation or condition over time. Development, however, implies change, usually an improvement of a situation or condition over time. Hence, sustainable development can be viewed as the maintenance of a positive rate of improvement.

Improvement involves change. But the improvement of societies or systems societies create can only be realized by adaptation to the processes of change, substitution and replacement. The continued existence of particular products or technologies or engineering structures or even social institutions is not a necessary condition for sustainable development. Quite the opposite. A sustainable economy can only be realized if there is continued adaptation, creation, and innovation, the implementation of new knowledge, new attitudes and new technologies and new operating policies to the betterment of humans and their environment. Water resource systems are no exception. The sustainability of water resource systems will almost always require periodic modification of those systems to meet changing demands and conditions. There is a difference between the sustainability of a particular component, like a reservoir, and the system itself. The former is usually impossible, the latter is possible if we have the wisdom to consider the long term impacts of what we do today, and take actions that will not preclude future generations from deriving the greatest benefits they can from their water resource systems.

One can contribute to sustainable development by designing and developing systems whose components can be changed or modified to meet changing demands or to take advantage of new technologies with relative ease and minimum cost. But as the systems are modified, some aspects of the original systems or the environmental impacts they created may disappear and be replaced by others. For example, a part of a natural ecosystem may be converted to economic activities (or vice versa), or one resource (e.g., fossil water from the desert aquifers in the Middle East) may be consumed to permit the production of another resource or even to contribute to something as intangible as new knowledge, technology and culture (as recorded in books, as implemented throughout society, and as shown in museums). Unfortunately, the appropriate tradeoffs, what non-renewable resources one should consume over space or time for the possibility or promise of something else that is supposedly better than what now exists, are difficult to determine. Broad interpretations of sustainable development make it very difficult to define precise measures of its attainment. They cannot help us identify working tools for determining progress toward achieving sustainability. This difficulty creates a need to develop a more operational or useful definition of sustainability and sustainable development, one specifically suited for water resources planning and management. Later chapters in this monograph address this need.

THE CHALLENGE OF SUSTAINABILITY

Since the Brundtland report of 1987 (WCED, 1987), sustainable development has become the focus of discussions and debates throughout the world. The potential regional and global impacts of climate change; stratospheric ozone depletion; forest, wetland and other natural habitat destruction; species extinction and accompanying loss of biodiversity; encroaching desertification; and contamination of aquifers and surface water supplies are all pointing to the conclusion that many present development and management strategies must be altered. Failing that, it may not be possible to continue to develop and prosper as we have in the past, or to the extent we have the potential to in the future.

The consideration of the impacts of what is done now – on present as well as on future generations – is the essence of the term sustainability. It is not a new idea and it is not a new concern. The renewed interest in this topic largely comes from an increasing public awareness of the potential threats to local and global environments and to the adverse consequences of those threats to the quality of life. How to allocate and manage available capital and human and natural resources in a manner that improves the quality of life now and in the future has become one of the major economic, environmental and social issues.

An explicit consideration of the needs or desires of future generations may require us to think about giving up some of what we could consume and enjoy in this generation. How should these tradeoffs among various generations be made? A standard economic approach to making these intergenerational tradeoffs involves the use of discounting, at some rate of interest, future benefits, costs and losses. This type of analysis requires converting all future benefits, costs and losses to equivalent present day values to account for inflation and the time-value of money. Simply put, most of us are willing to pay more for something we want today than for the promise that the same thing will be given to us at some future date. That 'thing' could be money. Money available today can be invested and earn interest over time. Hence a given amount of money invested today will increase in value over time. Future values, when discounted to the present (using a specific discount (interest) rate) will be worth less today.

Since future benefits, costs and losses, when discounted to the present time, decrease in value, any loss or cost that occurs in the future will have less effect on the present net benefits than will the same loss or cost occurring at the present time. Consider, then, the result of such a discounting procedure if the supply of a supporting environmental resource is limiting, such as the quality of the soil of an irrigation area or the assimilative capacity of a groundwater body. Should decisions made today allow this land or groundwater resource to be degraded in the future just because that future loss, when discounted to today's values, is nil? Most would argue that the concept of discounting in benefit–cost analyses is valid. But, they would generally agree that discounting should be applied with safeguards where the integrity of our life-supporting resources – such as fertile soils, potable water, clean air, biodiversity and other environmental and ecological systems – are concerned.

But how can any safeguards or constraints on the traditional benefit-cost analysis be applied, and by whom, to ensure that those who live and consume today will adequately consider the needs and desires of those who follow them? How can we possibly know what those future needs and desires are? To what extent do people understand what the impacts of their actions might be on those living fifty or a hundred years from now?

Over the past century resource economists have been telling us why it is so difficult to manage and use our environmental resources today in a way that will benefit those of us living today, let alone those who will be living in the distant future. Any study of recent history also shows how difficult it has been for governments to modify either a free-market system or a centrally-planned and controlled economy (by means of taxes and subsidies or by laws and regulations) in attempts to ensure any sustainable use of common property environmental resources. But these difficulties should not be excuses for ignoring sustainability issues. Rather, those who are managing natural resources need to ensure that the public and those who make decisions are aware of the temporal as well as spatial sustainability impacts and trade-offs associated with those decisions.

When considering tradeoffs of natural, capital and social resources that affect the welfare of humans and other living organisms over time, one must also address the question of spatial scale and resource mobility. Should each square kilometer of land be sustainable? Should each watershed or country or province or state be sustainable? Might some large regions (e.g., the Aral Sea and its basin) be rightfully sacrificed in order to enhance the economic survival of a larger region or country? Opportunities for resource transfers and tradeoffs and for the achievement of sustainability are generally greater the larger the space scale. However, concern only with the sustainability of larger regions could overlook the unique attributes of particular local economies, environments, ecosystems (and possible limits on ecosystem adaptation), resource substitution and human health.

Given these and no doubt many other questions and issues, it is clear why there has been so much difficulty achieving a consensus on just what is meant or implied by sustainability or sustainable development. Most will probably agree that sustainability involves an explicit focus on at least maintaining (if not increasing) the quality of life of all individuals over time. They will probably also agree that sustainability also addresses the challenge of developing regional economies that can ensure a desired and equitable standard of living for all those living in the region, now and on into the future. Probably few, however, have any idea as to how this will be done, or even if it can be done. Some will even argue that it cannot be done. It will, without question, require some truly interdisciplinary research over a considerable period of time to address and answer many questions and issues. Some of these include:

- How can water and other environmental resources be managed so as to achieve the highest possible quality of life for those living now and in the future?
- How can these resources be managed to increase the level of equity in the quality of life that exists among all populations in developing, redeveloping and developed regions?
- What is equitable?
- How will future generations continue to improve their standard of living?
- What will be the values, goals and expectations of those future generations?
- Who will have to pay, and when and how much will they have to pay, to achieve any degree of sustainability?

- When and how will sustainable development be achieved?
- Is such an achievement even possible and if so, how can we know we have achieved it?
- What should we be giving up or doing differently today to enhance, more equitably and more reliably, the quality of life of future generations?
- How can improved technology facilitate the achievement of sustainable water resource systems?
- How can the overall regional impacts of sustainability constraints on the design and/or operation of water resource systems be determined?
- Do such constraints imply rates of renewable resource consumption no greater than rates of replenishment. If so, over what time and space scales?

Similar questions apply with respect to the rates of waste generation and discharge compared to the rates of waste assimilation in water. Can measurements be made of the sustainable yield of a resource like water whose supply and quality are random (stochastic) over time? Droughts and floods of various magnitudes and duration will no doubt continue to occur, and global warming may actually result in increasing those magnitudes and duration. What statistical measures like reliability, resilience or vulnerability for various criteria could be used to measure long-term sustainability?

None of these questions are easy to answer. Answers to all of them will require more understanding. A better understanding will require more research. But at the same time the practice of water resources planning and management should not wait until we obtain this better understanding. There is an obligation to our descendants, if not to ourselves, to begin increasing the sustainability of water and other natural resources today.

It will be difficult to reach a state where an equitable and desired quality of life can be guaranteed. However, the fact that these issues are being discussed and that measures are being taken to achieve that state, indeed at the highest levels of many governments, is a start. And while such goals or states of equity may never be met, one can argue that they are goals or states that present and future generations should continually try to achieve. One fact is certain. General sustainable economic and social development will never be achieved without the achievement and support of sustainable water resource systems.

Water resource engineers and scientists have an opportunity to accept the challenge of leadership in the move toward sustainable development of these resources. Those who are responsible for developing and managing water and other environmental resources must be among the leaders in this transition toward a sustainable future. What we do with our natural environmental resources has substantial impacts on economic development and human welfare. Are the actions and decisions leading society toward one that can become sustainable? If not, or even if so, can a better job be done? And, if so, how? To begin to address these questions requires an understanding of what is implied by sustainability and how that concept can be made operational.

Water resources management is an important aspect of economic, environmental and social development. Water resources systems consist of natural ecosystems, engineered facilities, and management institutions. Everyone involved in those systems has an obligation to see that they can provide sufficient quantities and qualities, at acceptable prices and reliabilities, to meet demands now and in the future without causing environmental deterioration. Are they using appropriate technology that can be operated and maintained locally? Are all the costs of various structures, including their upstream and downstream impacts, and of their maintenance and replacement or disposal or abandonment included in their overall costs? Are all costs and benefits being allocated equitably? If these questions are not all answered, it is likely the particular water resource systems are not sustainable.

Figure 2.1 Water: the resource that makes our Earth unique among all other known planets in the universe.

WATER RESOURCES LIMITS TO SUSTAINABILITY

If there is one thing there is a lot of in this world, it is water. This can be seen from any picture of the earth from space, as shown in Figure 2.1. Those who estimate such statistics tell us there are about 1.4 billion (thousand million) km^3 (over 300 million cubic miles) of it. However, apparently only about 2.5 percent of all this water is fresh, and only about 0.3 to 0.4 percent of the fresh water is renewable. Most of this renewable fresh water evaporates or becomes lost to deep groundwater aquifers. Approximately 40,000 km^3 per year of renewable water becomes runoff. If this were divided equally among the some 5 to 6 billion people currently living on earth, each could consume an average of over 20 m^3 of this renewable fresh water per day. This is about an order of magnitude (10 times) more than the amount most of us currently need or use, let alone consume. This includes the amounts allocated to meet agricultural, industrial and instream demands.

From a global perspective, then, there is no water shortage. At least not yet. Perhaps this is why those who are

mainly concerned about the global dimensions of sustainable development have tended not to give very much attention to water. On a planet whose surface is more than two-thirds water, the illusion of abundance can obscure the reality that renewable and useable fresh water is an increasingly scarce commodity for an increasing number of people.

But everyone knows that water is not distributed evenly over the earth. Nor is it distributed in accordance with the earth's populations and their demands. On a local and even a regional scale, water taps (where they exist) can run dry. This has occurred in many places. It can be the result of quite different circumstances. In less prosperous regions there may not be any taps to go dry. Who knows how many millions (mostly women) walk how far each day to find and carry the water they and their families use and consume? And the amounts each of them carry are far less than the worldwide average per capita use per day.

In many regions where populations are growing most rapidly, the needed water is simply not available. As seen in Figure 2.2, this is especially true in much of Africa and the Middle East, and now increasingly in parts of Asia, Latin America, and southern Europe as well. Critical limits of useable water are not at the global level. They are at regional and local levels. (Some of the important research needs at these levels have been identified in a report of the US National Research Council, 1991a.)

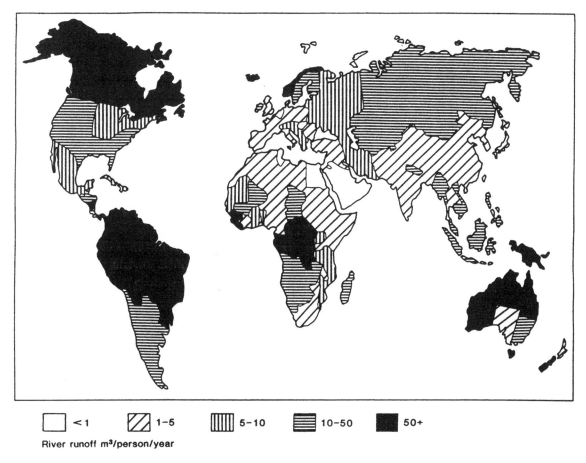

Figure 2.2 Ranges of river runoff per capita per year as estimated by L'Vovich (1977).

FINANCIAL LIMITS TO SUSTAINABILITY

A major constraint on the development of sustainable water resources projects will always be financial. Obviously a large-scale development project cannot be effective if the financing is insufficient. Financing is required not only for construction, maintenance and operation, but also for planning and operation studies. Both time and finances have to be available to produce a properly designed plan based on sufficient data. Unfortunately, many plans have had to be developed based on poor economic, hydrologic or climatic data. The amount of information desired will always be more than that available, especially with respect to the future. Under such conditions it is even more important to design very robust systems that can adapt to a range of plausible futures without undue costs. Whether or not the results will be sustainable may not be known until after some years of project operation, monitoring and modifications based on the results of the monitoring.

Operation of any water resource system requires continual financial support. Project planning processes must, therefore, include identifying or developing ways of financing the management and operation of the project. The sustainable operation of a system depends on properly functioning infrastructures and the provision of funds for maintenance and personnel.

Often the ways in which services are priced by publicly operated water resource systems have little to do with actual costs and benefits. Many economists recommend that private enterprises should plan and manage water development projects. Throughout the world many projects and systems are developed and managed by private companies, companies that have economic efficiency as at least one of their goals. The involvement of the private sector, the argument goes, leads to more efficient operation and services because there exist financial as well as social incentives to improve the performance of the system and its personnel. Clearly evidence can be found to support this argument. On the other hand, in regions where private water management is not feasible for whatever reasons, there is no reason why public agencies cannot provide effective and efficient services whose costs are recovered. One way of obtaining the necessary funds is through proper water pricing. If socially feasible, one-time capital costs and recurring operation and maintenance expenses should be introduced on a 'users pay their fair share' principle. Many different cost distribution methods, like staggered price controls, exist. Similar considerations based on the 'user pays' principle could also be applied to other activities involving the development and use of natural resources. Regardless of whether the system is public or private, unless the costs of providing the services demanded from any water resource system are recoverable, such systems cannot be sustainable.

TIME AND SPACE SCALE ISSUES

Any definition of sustainability involving the development and management of a resource having a variable and uncertain natural supply, such as water, must take this variability into account. Whether or not a system is considered sustainable will depend on the length of the time period and on the size of the region being considered, the region's population diversity and density, and its various cultural, economic and social values and standards. Sustainable development does not imply that the quality of life (however measured) of each person or the quality of the environment or ecosystem of each square meter of land must improve in each successive minute, day, month, or even year. However, what space and time scales are appropriate?

All water resource systems will certainly experience events such as floods and droughts sometime in the future that will lower, at least temporarily, the quality of life of those who are impacted. Sustainable development criteria for water resource systems must therefore somehow take into account the natural fluctuations in supplies and demands – fluctuations that can, on occasions, result in decreases in the quality of life for those using or being impacted by such systems. Again, what space and time scales are appropriate?

There is no doubt that the perception of sustainability, and thus the motivation for action, depends on an ability to understand the interactions of processes on very different scales. The scale that is the easiest to perceive is the local scale at which a person can see and understand dependencies. Somewhat more abstract is the regional scale, as exemplified by the combination of city and surroundings or by combinations of small river basins. The least comprehensible scale is the global scale of international or even larger interactions that an individual can only understand in an abstract way. It is usually so far outside an individual's experience that the connections between actions and their effects cannot be seen.

Until recently, the degradation and depletion as well as restoration and replenishment of environmental resources have been observed mostly at local or regional scales. In the past decade, however, we have learned that human economic activities can also adversely and possibly irreversibly impact our global environment. The survival of human civilization will ultimately be determined by the ability of people to master sustainability on a global as well as a local and regional scale.

Globally environmental impacts are often characterized by small interactions occurring over large distances. The effects of those impacts can be detected only in world statistics, and may be overlain by other influences. For some years now measurements have shown a general deterioration in global environmental resources. Every concerned individual has seen statistics on:

- inefficient energy production and consumption,
- air, soil and water pollution and its adverse economic and health effects,
- the increase of greenhouse gasses in the atmosphere and the resulting increases in global temperature and sea levels,
- ozone depletion in the upper part of the atmosphere and ozone accumulation in the lower part, both adversely affecting animal, plant and human health, and
- deforestation and desertification.

Of course, there are more.

Does the burning of fossil fuels in Europe contribute to water shortage in Africa? Till now, the reports of increasing vulnerability of distant regions, or of long-term changes in environment, have had little influence on individual life styles elsewhere. But consider, for example, the state of the water quality of the major seas of Europe: Baltic Sea, Black Sea, Mediterranean Sea and the North Sea. Although large parts of the populations of countries whose rivers flow into these seas, or of countries that border those seas, are concerned about environmental degradation and the loss of fish stocks, the actions required to reduce the input of pollutants and to protect those seas involve difficult national as well as international issues and are progressing only slowly.

On local and regional scales, engineering theory and experience have yielded solutions to most water needs such as water supply and sanitation, irrigated agriculture, flood protection, energy production and, to some extent, transportation. Conventionally, a sectorial approach to water resources development has been, and continues to be, used. Local and regional needs are identified by appropriate decision-makers. Engineers and other professionals are then typically asked to design and build systems that can meet those *needs* or *requirements*. Single and multiple-purpose projects are designed and built depending on the particular needs and on the available water resources and funding. They are then operated to meet those needs and to reduce any negative impacts resulting from such development. Cost is the usual constraint to unlimited, or even economically optimum, development.

This local and regional *meet the requirements* approach often fails to consider sustainability. That is unfortunate. At some point professionals must get themselves more deeply involved in the process that generates these *requirements* and *needs* that engineers are told must be met and that almost always seem to require development of more and more water resources infrastructure. Professionals must also be involved in the *management of demand*. They should also enter the debate on what level of development is appropriate. Since there is not much money to be made arguing for limiting development, this will not be easy for consultants. Yet many professional societies and organizations have established sustainability guidelines that can help all of those involved in this debate (see for example da Cunha, 1989; German Ministry for the Environment, 1994; Institution of Engineers, 1989; World Bank, 1993, 1994; Gleick *et al.*, 1995; Hufschmidt & Tejwani, 1993; Pennekamp & Wesseling, 1993; United Nations, 1991; van den Bergh & van der Straaten, 1994, to mention only a few).

WATER SUPPLY AND DEMAND MANAGEMENT AND POPULATION GROWTH ISSUES ————

As far as anyone can tell, the demand for water has continually increased over time. There are two reasons for this. The first is that populations have increased, and as populations increase, more and more water is demanded for domestic, industrial, agricultural and other purposes. The second reason is that average standards of living have also tended to increase. As standards of living increase, so do per-capita water demands (Engelman & LeRoy, 1993).

Demand management addresses ways in which water is used and the various methods available to promote more desirable levels and patterns of use. Water conservation (use reduction) measures and water pricing policies are two instruments of water demand management. Effective demand as well as supply management are equally important tools for any sustainable water supply and distribution system.

In many regions of the world, the demands for water have been increasing to the extent that they have, or soon will,

exceed the renewable water supplies that can be collected, treated and distributed economically. So far, nature and developments in engineering technology have not kept up with that increasing demand.

There are a number of ways to augment existing fresh water supplies. One option is desalination of sea water and the subsequent transport of the desalinated water to the water users. In many places where desalination is a technically feasible option, it still costs too much to desalinate water from the sea and pump it to where it may be needed. The costs of desalinated sea water are likely to decrease over time, but in the foreseeable future it will still cost more to produce it and transport it than most are able or willing to pay.

Another option for augmenting water supplies is the transport of water over long distances, over land in pipelines (such as is currently practiced in Libya) or canals (as in many irrigation areas), or by sea in tankers or in floating flexible bags. Tangier was supplied with fresh water in the summer of 1995 by tankers that brought water from further south on the Moroccan coast. Large-scale water transfers from regions of 'excess' water to those not having enough, while technically possible, generally bring with them unacceptable environmental, economic and political costs. Proposals for hauling water in flexible bags from water-rich regions to water-poor regions along ocean or sea coasts have been promoted for some time. They have been considered by some countries bordering the Mediterranean Sea and along the west coast of the USA, among other places. While no major feasibility project has been implemented, this may prove to be a viable option in some regions of the world and may actually be implemented in the not too distant future. The transport and use of icebergs have so far not proven to be of much value for augmenting fresh water supplies.

Many arid regions are witnessing large increases in population and in per-capita water consumption. These increases are made possible mainly by withdrawals of non-renewable ground waters from aquifers. While some of these aquifers contain substantial amounts of groundwater, some do not. In any case economic and population growth based on non-renewable water supplies cannot be sustained beyond the limits of the supplies. Unless there is some technological breakthrough, it is doubtful that even current economies and living standards (or even populations) of some of these arid regions can be maintained. Legal constraints and pre-defined allocation and operating policies that do not consider all beneficial uses and states of a system also can be major factors preventing what might otherwise be considered socially effective, efficient, and sustainable water use allocations.

Treatment and reuse of sewage waters is becoming a common source for additional water in some water scarce regions. Reuse of sewage waters, when properly managed, has the benefit of reducing environmental degradation. As domestic demands rise, so does the quantity of sewage produced by the cities. These cities should be charged with collection, treatment and disposal of the treated effluents, in a manner which protects human health, water resources, and the environment. Reuse, primarily for irrigation, may be an economically attractive alternative. Who pays what portions of the total water reuse cost? Reuse can reduce treatment costs if what is being removed by treatment (e.g. nitrogen) is actually desired by those using reused water (e.g. irrigators). In several arid and semi-arid regions of the world, reclaimed sewage is the next source of 'additional water' that replaces fresh water taken away to meet increasing domestic water consumption demands. Desalinated sea water is usually much more expensive.

Two main obstacles to the use of reused water are health fears, and the added cost of dual water supply systems that may be required. Often putting treated effluents into surface or groundwater bodies and then withdrawing that same water from those water bodies is publicly acceptable even though direct reuse is not. It is one method of creating 'fresh' water from previously used and treated wastewater.

WATER QUALITY AND HEALTH ISSUES ——

Water is of little (or certainly of greatly reduced) value if its quality is so poor that it precludes many of its potentially beneficial uses. Considerable sums have been spent on water and wastewater treatment in both the developing and developed regions of the world. Yet statistics indicate that in spite of considerable amounts of these investments in water quality improvement and protection, they represent only a fraction of the investments needed to substantially reduce waterborne diseases and meet commonly accepted environmental and ecological objectives (UNESCO, 1992).

Wastewater management methods should be appropriate to the region and its culture and economy. In many developing regions, for example, night soil removal and disposal is much more cost effective than sanitary sewer systems and wastewater treatment. In some regions western technology can be a more expensive and less reliable way to control pollution from human domestic and industrial activities.

For some arid and semi-arid regions, re-use of wastewater may contribute more toward future water availability than any other technological means of 'increasing' water supplies. Treated wastewater can be used effectively for irrigation, industrial purposes, and groundwater recharge and for protection against salinity intrusion in groundwater aquifers. In many regions industrial demands are being met increasingly by treated and recycled waste waters.

Poor water quality is most costly in the developing world (United Nations, 1992). The results from the recent International Drinking Water and Sanitation Decade (Traore, 1992) showed that money and technology alone are not going to improve water quality, sanitation and public health. Actions contributing toward a more sustainable water quality management program will include:

- population-demand control,
- institutional reform,
- implementing incentives for improved cost effectiveness,
- more training at all levels of society on the benefits derived from better water quality and waste management practices,
- training in the operation and maintenance of water and wastewater treatment facilities,
- implementation of cost recovery options,
- improved monitoring,
- non-point pollution control and greater recycling of waste residuals.

Most important among all water supplies is that used for drinking purposes. Drinking water supply development typically includes the traditional activities – locating and developing new sources of water, treating the water as necessary, and transporting and distributing that water in a cost effective way. Water distribution systems can range from public wells or springs and street vendors to elaborate city distribution systems involving extensive networks of pipes that are supplied through pumping stations and temporary storage in water tanks, as required.

Drinking water can also be obtained by desalinating sea and brackish ground water. But this technology is relatively expensive. Its high investment cost will be overcome only when the requisite equipment can be produced in sufficiently large quantities, and when the high energy consumption of the desalination process can be reduced. For the moment, the energy requirement with different technologies renders this technology feasible only where energy is cheap and water extremely scarce, e.g., in arid oil and gas producing countries.

In the less developed countries the lack of adequate safe drinking water supplies and sanitation facilities is still the situation facing nearly 2 billion people. This is the case even after a decade of attention and financial aid from national and international agencies that was specifically aimed at eliminating these conditions (Okun, 1991). The situation is especially critical in urban areas. During the second half of the 20th century, while the world population will have grown by 150 percent, the urban population growth will have grown by 300 percent. By the end of this century, the United Nations (1989) estimates that almost half of the world's population will be living in cities.

The growth of mega-cities places a considerable stress on drinking water supply systems. Historically, civilizations developed along major rivers, and most large cities today are in such locations. In many such cities the water supplies available from rivers or local groundwater sources are no longer sufficient, mostly because of their poor quality. Hence, these cities are increasingly being supplied with water from distant reservoirs or aquifers. This water is transported by aqueducts and tunnels often over considerable distances. In both developing and developed regions, large-scale water projects designed to enhance economic development and human welfare have been widely promoted and subsidized since the end of World War II.

Many parts of the world are increasingly considering a bottom-up approach to domestic water supply policy definition and implementation. While strongly influenced by financing agencies' policies, there is a growing recognition that top-down management plans do not always work well, and that increased attention must be given to user-driven directions – emphasizing cost-recovery, financial management, and operation and management of the constructed systems. There is little doubt but that there is, and will continue to be, a growing emphasis on giving water user groups more responsibility in the management of the systems.

Water quality problems are certainly not restricted to urban areas. The lack of sanitation facilities and the too often associated unsafe drinking waters remain among the principal causes of disease and death, especially in rural areas. Specific measures to counter water-related threats are often needed but lack of money and inadequate local management often lowers their effectiveness.

GROUNDWATER MANAGEMENT AND USE ISSUES

For domestic water supply and in many cases also for agriculture (particularly for high value cash crops) current tendencies are toward more extensive use of ground water.

Today in many semi-arid countries most of the supplies used to meet urban and domestic water demands come from ground water. In many developed regions surface waters are often not clean enough to be used directly. Hence water treatment is necessary to remove pollutants, disease vectors and other impurities. It is often less expensive to use ground water. It is for these reasons, in part, that ground water is currently the primary source of drinking water even in regions with substantial surface water supplies, such as in Germany, for example.

Relatively recently, ways have been explored to augment the groundwater reserves by infiltration of excess surface waters from floods, or by infiltration of treated waste water.

As noted before, if groundwater reserves are not being recharged at rates equal to or greater than the withdrawals, as is the case in many arid regions, then these withdrawals cannot be sustained. This does not mean there should be no withdrawals from aquifers that have little or no recharge, but rather that some analyses need to be performed to determine just how much should be withdrawn and when. Such analyses are discussed in more detail in later chapters. A self-regulating mechanism against excessive withdrawal of ground water arises from the cost involved in pumping from increasingly deeper groundwater reserves. In some areas these expenses have led to the abandonment of farms that were once irrigated with ground water that had become increasingly expensive. An example is the use of the Ogallala aquifer in the Middle West of the USA (which will be discussed in more detail in Chapter 5).

Ground waters of many regions are increasingly being threatened by pollution and encroachment of urban settlements. When the area above an aquifer is used for agriculture, industry or urban development, pollutants from these activities will find their way to the ground water. While pollutant infiltration is generally slow, when aquifer pollution does get noticed, it is very costly to remedy. Perhaps because contamination of aquifers is not noticed immediately when pollutant discharges occur, groundwater protection is usually only a secondary issue in decisions regarding land use in highly populated and industrialized countries. Yet, groundwater protection is often vital to the other interests of infrastructure, industry and agriculture. Extensive protective actions are sometimes needed to preserve groundwater quality. Improved land-use management, needed to increase recharge by reducing runoff and evaporation and to prevent contamination, is an important consideration wherever ground water is used, especially in arid climates.

Other challenges to sustainable groundwater management can arise from the difficulty of locating and assessing the yield of groundwater reserves. Poor assessment of groundwater resources, inadequate analyses of the consequences of groundwater exploitation, or lack of technical understanding and inadequate institutional frameworks, may lead to management decisions that could cause an irreversible loss of ground water or even the aquifer storage capacity.

The conjunctive use of surface and ground waters can sometimes increase the sustained use of both water systems. For example, water management in parts of Europe is based on the use of ground water as the basic supply, augmented by water from one or more surface water reservoirs in times of large demand. Water from polluted rivers is also cleaned by means of bank filtration, or artificial recharge of ground water. In other parts of the world, where high evaporation rates make it prudent to use surface waters as quickly as possible, ground waters are used as the emergency supply in cases where surface waters cannot meet the total demands. Sophisticated management strategies are used, for example in Israel, to maximize the amount of water available through an efficient combination of infiltrating surface water and pumping of ground water. Quite often, however, groundwater pumping is in excess of recharge. This can result in sea water intrusion along coastal aquifers and in land subsidence with accompanying damage to surface and subsurface infrastructure.

Clearly for all the reasons cited above, the sustainable management of groundwater resources poses one of the most difficult and complex problems of water resource systems management.

ENVIRONMENTAL PROTECTION ISSUES

An inseparable part of water resources management is environmental protection against pollutants that threaten the quality of water supplies. In many developed countries, one of the costliest tasks is reclaiming contaminated sites. Such conditions might have been caused by past mining or industrial activities, or they might be the result of excessive salinity from poor irrigation practices. If possible, the costs of remedial measures should be charged to the polluter ('polluter pays' principle), but in many cases the polluter of ground or surface waters is no longer around, or is financially not able to pay for cleaning up those water bodies. Nevertheless, a guiding principle for all future activities that may lead to contamination of soil or water is that the cost of cleanup be paid by those who contaminate.

If those living now must pay for the required remedial measures of any contamination that they produce, they will

be less likely to produce it. This is easier said than implemented in practice, however. In many cases where pollutants from multiple sources are discharged into water bodies, it is difficult to identify, and prove in court, whose pollutant is violating a quality standard. This would be less of a problem if potential pollutants were tagged at the time of manufacture.

The effective disposal of sewage and other waste waters has been a challenge to environmental and public health engineers for more than a century. Wastewater disposal is often a major factor in the health of a population. In developed regions, sewage from domestic sources (collected in closed conduit sewer systems) needs large quantities of water to transport the wastes. In many developing regions the sewers are often open man-made or natural channels, if they exist at all. Alternatively, undiluted sewage is collected as night soil. At locations where dilution of the sewage is not feasible, sewage disposal plants have sometimes been designed, constructed and operated – but often not too effectively.

Wherever possible the prevention of pollution is preferable to the reduction or elimination of its consequences. This commonly involves eliminating potentially polluting activities, treating the remaining point sources and implementing effective measures for reducing non-point source pollution originating from urban and agricultural lands in the watersheds. Non-point source pollution control is particularly important for the protection of ground water. Once polluted, ground water aquifers cannot always be cleaned without substantial investments of time and money.

For both chemical and biological pollution of near surface ground waters, natural ecosystems (e.g., wetlands) are increasingly being proposed as a means to partially reduce the levels of pollution and to increase the sustainability of pollution control. Significant scientific input is needed to obtain an adequate understanding of the socio-economic and ecological processes when developing effective ecosystem management strategies and policies.

HYDROELECTRIC ENERGY ISSUES

Hydropower is a well-established technology and is one of the cleanest sources of energy available. Where abundant water resources and varying topography exist, it can be a major source of power and energy. The Pacific Northwest of the USA is one such region. The Tennessee Valley in Southeastern USA is another. Norway's energy production is solely from hydropower. Large-scale developments can provide large, concentrated quantities of electricity needed to meet industrial, commercial, and residential demands. Small-scale and micro hydroelectric plants can provide power to isolated, sparsely populated communities and agricultural processing plants. Unfortunately, hydropower can only provide a small fraction of the energy requirements of most developed energy-intensive regions even if innovative approaches are followed.

The number of hydroelectric pumped-storage plants is increasing rapidly. Pumped storage, where cheap energy is used to pump water up into storage reservoirs for later use in generating hydroelectric energy when the energy price is higher, is one of the few ways to store electricity for use in peak demand periods. Most importantly, hydro and pumped storage provide operational and planning flexibility to electric utility systems. In systems that generate electrical energy mostly from thermal energy sources it is almost impossible to determine the actual value of hydro generation and pumped storage because they indirectly make it possible to operate thermal sources efficiently.

Many countries, particularly developing countries, are installing small plants (of 15 MW or less) at village levels. These local decentralized power stations often provide a more sustainable, and a more reliable, source of energy. (In developed regions, however, such local hydropower plants do not usually lead to reductions in central system capacity due to their relative low reliabilities.) Apart from economic benefits, however, these installations have contributed to the quality of rural life. Small-scale hydropower is often able to use indigenous labor and materials. China's experience in developing appropriate indigenous approaches with its own capital and technology is unique. It has resulted in tens of thousands of small hydro-stations. Because of their proven advantages and benefits, small hydroelectric plants are likely to receive increasing attention in regions where they are feasible.

AGRICULTURAL AND INDUSTRIAL DEMAND ISSUES

Irrigated agricultural and many major industrial plants are among the largest users of water resources throughout the world, and hence deserve some special attention here. Their all-too-often inefficient use of limited regional water supplies is a major threat to the sustainability of those water supplies. Ideally, water use allocations should be competitive, made only after all demands for water have been considered, and not just those of agricultural production or industry. These include domestic/municipal, industrial, agricultural and instream ecosystem requirements and demands.

It is particularly vital that irrigation system planning include the drainage required to avoid salinity and water logging problems in irrigated areas. It is also vital that the water distribution and drainage structures associated with irrigation systems be properly maintained. If these concepts are respected, irrigation systems can be and have been sustained over long periods. Numerous examples of well functioning irrigation systems, in which operation and maintenance work hand in hand, exist throughout the world. Counter-examples include irrigation systems destroyed by siltation of the reservoir supplying the system, by salinization of the soil usually from faulty application of well-known rules, from negligence, from over fertilization and subsequent algae growth, or from incomplete development, e.g., ignoring drainage or canal leakage and the consequent rise in groundwater levels in order to save on capital investment.

Irrigation can lead to a number of detrimental consequences, particularly changes in water quality and water logging. Drainage water from irrigated fields and surface runoff are carriers of excess fertilizers and pesticides. These constituents can adversely affect the reuse of returned drainage flows downstream. Poor irrigation practices can also create habitats favorable to carriers of water-borne diseases, such as snails and mosquitoes. In addition, in closed irrigation systems where evaporation is the method of disposing of excess waters, residual heavy metals leached from the soil can create hazards to wildlife.

The non-point pollution input to surface water systems from agricultural fields is often increased by substances attached to the soil that are removed through surface erosion. This occurrence is another reason for preventing the loss of fertile soil due to erosion by wind, rain or snow melt. One way to reduce soil loss in the agricultural sector is to prevent overgrazing or deforestation. Soil erosion not only leads to the loss of valuable topsoil but also causes silting, sedimentation and turbidity in downstream areas. Although storage reservoirs will always receive some sediment, the upper watersheds can be managed in ways that minimize reservoir sedimentation. Erosion should be prevented primarily by eliminating its causes; however, in many cases this is a task of formidable proportions. For example, in many tropical regions deforestation of rainforests is taking place to the extent that it has become the most important environmental issue in those regions. We have known the consequences of such actions for a long time.

Consider, for example the following story taken from the writings of E.W.D. Jackson as reported in UNESCO (1991). A 'Lesson' from the Past:

In 1935, Mr. E.W.D. Jackson, a British civil servant working in Burma, warned of the accumulation of sediment in Meiktila Lake, an irrigation system with a unique history. Writing in *Irrigation and Drainage Paper Special Issue 17*, Jackson stated, "Meiktila Lake is without doubt the most successful (and sustainable) irrigation work constructed in the time of the Burmese Kings, and remains substantially unaltered since the day it was completed some 900 years ago." Jackson pointed out that the lake, located in central Burma on the slopes of Mt. Popa, had been more than just an irrigation work. It was a historic monument "regarded with feelings of pride and deep sentiment by every Burman," However, by 1935, Meiktila had begun to silt up. In his paper, Jackson pointed out why:

In the time of the Burmese Kings there was no such trouble. The lake was famed for its clear water, in which even weeds did not grow. The bunds (embankments) were never in danger of breaching – the only recorded breach was due to a cut made by a Chinese army that invaded in the 15th Century.

If the lake could last nearly 900 years under Burmese rule without silting up, what change of conditions has occurred under British rule to [have] placed the lake in such grave danger within recent years?

In the time of the Burmese Kings there were very strict laws enforced in connection with the administration and control of the Meiktila Lake. One of these laws in particular was enacted with the dual intention of ensuring a good regular supply of water and preventing the entrance of silt into the lake.

No person, on pain of death, was permitted to cut any jungle or to clear any land for any purpose whatsoever within two miles of the bank of a stream in the catchment area.

The Burmese knew that in closely afforested areas there were usually ample supplies of subsoil water, which was of course very desirable in the streams feeding the Meiktila Lake.

It was also a matter of common knowledge, based on ordinary observation, that the protection of wide strips of land by forests, undisturbed jungle, and grass would prevent eroded soil from being swept into the streams and carried down as silt during the rainy season.

(However), under the British administration many harshly enforced Burmese laws became obsolete, and unfortunately the laws concerning the Meiktila Lake were amongst them.

The engineers were given the lake itself to look after and improve. The catchment area was not placed under the control of the irrigation authorities (who, it must be confessed, did not in the light of the technical knowledge of those days, concern themselves very greatly about what might occur in the areas above the lake), but was thrown open to cultivation to a great extent, and in parts worked as a reserve forest for purpose of revenue.

The natural result has been that the land in the catchment area has been cleared, especially the rich strips of virgin soil along the banks of streams, and the effect of the climate has accelerated surface erosion to such an extent that the situation is now one of the utmost gravity.

TRANSPORT AND FLOOD PROTECTION ISSUES

Inland navigation remains one of the most cost-effective ways of transporting goods. Among the conventional modes of transport (rail, road, air and water), the waterway has the least impact on the environment. It is also an economical means of bulk transport over long distances, as it uses the least energy per ton of freight transported. To make rivers navigable for large ships, however, river training along with periodic dredging and navigation locks must normally be provided. River training works are designed to reduce bank erosion and damage of the shore from ship waves. These works can also protect against damage from floods, snow melt or heavy rainfall.

Interference by humans on the behavior of a river, nevertheless, invariably brings about permanent changes of the river's course. In Europe, for example, the training of the Rhine and of the Danube has changed these rivers into a staircase system of dams and locks. In 1992, the completion of the Rhine Main Danube canal provided a connection between the North and Black seas. Training of the Wisconsin River (one of the busiest rivers in the US, for its modest size), the lower Mississippi River and the Tennessee River are examples of training works carried out on other major rivers used for navigation.

Technical river training measures can have detrimental as well as beneficial effects. These may include reductions in river bank groundwater elevations and induced changes of fauna and flora near the rivers. Often there are adverse visual impacts as well. Because of these adverse impacts, modern river training has become a multi-objective planning issue – in which the demands for shore protection, protection against floods, hydropower generation and navigation are combined with ecological considerations. River training works installed in part for flood protection can be damaged given a severe enough flood. Those who develop flood plains 'protected' by flood protection works are at risk, as known by anyone observing the recent floods in China, North America, and Western Europe.

Modern approaches are used in an increasing number of countries, including Germany and The Netherlands, where comprehensive management plans are set up for the large rivers. For example, there exists an 'Integrated Rhine Program' in which all the tasks required for use of the Rhine as a major navigable river are combined within a comprehensive plan. This includes flood protection of the cities in the Rhine valley between Basel and Mannheim, the protection of the remnants of the once abundant riparian forest and the planned reintroduction and maintenance of salmon in the entire river by the year 2000 (International Commission for the Protection of the Rhine, 1994). This course of action requires a detailed study of the river, the soil and the vegetation in the fluvial region, and the design and operation of structures such that they will not only impair the river flow as little as possible but also meet the combined objectives of esthetics, flood-plain ecosystem enhancement, cost and maintainability. The end result is expected to be much more sustainable than the largely engineered river system that exists today (van Dijk et al., 1994; van Dijk, Marteijn & Schulte-Wulwer-Leidig, 1995; Saeijs, van Westen & Winnubst, 1995).

RESERVOIR STORAGE AND OPERATION ISSUES

Surface and groundwater reservoirs are the primary means available to convert the natural spatial and temporal distributions of water to the desired spatial and temporal distributions. Besides converting precipitation to reliable yields of water supplies, reservoirs can provide storage volume heads needed for hydroelectric power generation. Reservoirs can also be operated so as to provide water-based recreational benefits and space for potential flood waters. A sequence of dams and associated locks can also provide adequate waterway depths for navigation.

While surface-water reservoirs can provide many benefits, they can also be the cause of many adverse environmental and social impacts, both upstream and downstream of the dam (see, for example, the assessment of a reservoir development project in India by Morse & Berger, 1992). There is no doubt that on one side of the balance sheet, such projects have sometimes meant ecologically stressful situations for animal and plant species, the forced removal of local peoples, deforestation and flooding of land areas, and other environ-

mental problems. Problems ranging from the loss of fertile soils, increases of greenhouse gas emissions, creation of environments for diseases and even local (and regional) climate changes have been claimed. No doubt under specific conditions, and without properly guarding against them during the planning stages, these events might occur. But, there are many benefits as well as disadvantages in the creation of reservoirs. Water is valuable and often a major key to social and economic improvement. Reservoirs can help to open up land for irrigated agriculture and settlement, and can provide the bases for industries and populations that are dependent upon a stable supply of water and energy.

To increase the sustainability of surface-water reservoirs, and indeed of the entire river system, it is necessary early in the planning stage to identify and carefully study all the beneficial and adverse ecological, economic, environmental and social effects associated with any proposed reservoir project. It is especially important to consider the long-term impacts of these structures. Future generations may have differing demands and goals. Hence reservoirs built today should be sufficiently robust in their design and operation to enable them to adapt to these possible changes.

NATURAL DISASTER ISSUES

A natural event that is unusual, such as an extremely rare flood, drought, earthquake, volcanic eruption, tropical cyclone, etc. becomes a disaster usually only if the people that live in the area where the event strikes are unprepared to cope with it. Very extreme events can cause large economic losses. They may disrupt social structures and impair development for years or even generations. The exchange of existing technologies and measures for disaster prevention among nations is a task that the international community of nations has begun to do, especially during the current United Nations led International Decade for Natural Disaster Reduction. Technologies and institutional mechanisms are available and must be used to prevent unusual natural events from becoming natural disasters

Water resources professionals have a responsibility to make this decade and future decades safer for people that are threatened by water-related extreme events. If countries have well-developed disaster management plans, damage from extreme natural events, such as the large scale floods in Bangladesh (1991) and in China (1990, 1996), can be reduced.

Water-related natural hazards that can become natural disasters include floods, droughts, land erosion, landslides and mud flows. Examples of human-induced hazards that can become human-induced disasters include dam breaks, levee failures, shipping accidents, accidental spills of toxic materials and structural or operation failures of water supply and wastewater treatment systems.

NATURAL AND MAN-MADE ENVIRONMENTS ISSUES

A guiding principle for the planning of sustainable systems is the concept that a water resource system should interfere as little as possible with the proper functioning of natural life cycles. Environmental impact assessments today are directed at finding out what impacts alternative projects will or may have on the environment and natural ecosystem and then applying a set of value criteria to rank these alternatives. It requires the consideration of alternate environments that can coexist with the water resources system, and deciding on the one that is most acceptable to the environmental as well as the social and economic sensibilities of the people who are affected.

A typical example is the creation of an environment for irrigated agriculture downstream from a reservoir. This can profoundly change the natural conditions that existed before the system was built. A sustainable development in such a case would require that the man-made environment be compatible with the soil and climate conditions of the region, and that the people affected have adequate opportunities to reap the benefits.

This aspect of sustainability is almost self-evident. However, it may require a new understanding of the interactions between humans and nature. For example, before the middle of this century German engineers straightened many of their smaller rivers in order to gain land for agricultural production. With agriculture no longer a primary concern, many of these rivers are now being changed back to more natural conditions. In fact, in the hydraulic engineering institutes of a number of technical universities and research institutions in Austria, Germany and The Netherlands the hydraulic conditions of planned 'renaturalized' rivers are being investigated routinely. Flood protection is now very often being accomplished by establishing more natural and ecologically productive areas in the flood plain. Hydrologists and hydraulic experts have cooperated with ecologists in planning for such areas in extensive portions of the Rhine and Danube basins. But in all such projects one runs into a problem: the definition of what is natural.

The University of Karlsruhe and the State of Baden-Wurttemberg are addressing this question of what is natural. They are defining 'Leitbilder' or model images of what a river should be like, based on reaches of rivers that are in ecological equilibrium. The 'ideal landscape' found in this way will normally not be achievable, it represents only an idealistic goal toward which planning is oriented (Larsen, 1992). For such work it is important to find out what sustainable and self-preserving natural conditions can exist under given hydrological and environmental constraints. One can then strive to obtain the most acceptable conditions, and to define criteria for this based on fundamental knowledge about stable ecological systems. The development and use of eco-hydrological methods helps quantity the interaction of hydrology (including the dynamics of the local water cycle) with the flora and fauna of a region. Such models are used to predict the changes in a natural ecological system if the hydrology changes, or the change in hydrology that would be required to create a certain stable ecological system.

CAPACITY BUILDING ISSUES

Sustainable system design and operation strongly depends on the capacity building infrastructure available in the region where it is located. Knowledgeable, well-trained and motivated individuals are the backbone of sustainable development. The people who can best manage a system effectively and efficiently are those who know what a system is supposed to accomplish and understand how it functions. Only those who are dedicated and meticulous and know what to look for will be able to detect and prevent the beginnings of deterioration, and thus properly maintain a water resources system. These are the reasons why it is so important that a capacity for infrastructure development and use exist in the regions containing the water resource systems. 'Capacity building' is one of the most essential and important long-term conditions required for sustainable development.

Another important factor in sustainable water resources development is that the local people must not only be capable, but must also be willing to assume the responsibility for their water resources systems. One of the drawbacks of a centralized dominating government that takes the responsibility for local system design and operation is that the local people become accustomed to looking to 'government' for help, rather than to looking to themselves. The ideal local water resources managers are well-trained persons who know the behavior of that system, have experience with its floods and its droughts, and know the concerns and customs of the people of the region, a group to which they belong.

Sustainable systems development and evolution cannot be achieved without local expertise, an expertise that needs to be developed and to be transferred to each succeeding generation of professionals.

THE INEVITABILITY OF CHANGE

An essential aspect in the planning, design, and management of sustainable systems is the anticipation of change: changes in the system components due to aging, changes in the demands or desires of society, and changes in the supply of water. These changes must be anticipated to the extent possible. The development of sustainable water resource systems must include planning for potential changes: changes in infrastructure to meet changing conditions in the natural watersheds and the changing demands by those who can benefit from the available water quantities and qualities. Sustainable systems must by necessity adapt to these changes.

Water resources planners and engineers must understand that the infrastructure they design, develop and operate will not last forever. They should therefore consider the impacts of its removal and replacement. While components of a water resource system are functional, they should function well even while other components, and even system objectives and expectations of system performance, are changing. For example, changing water uses, changing weather patterns or climate, and changing land uses, whether temporary or permanent, may alter the purpose and/or the performance of a system. A formerly forested catchment may be converted into agricultural land, or vice versa, altering its hydrological rainfall–runoff–groundwater recharge characteristics. Changes in runoff, erosion and sediment discharges into water bodies may occur. Surface water reservoir capacity reductions due to sedimentation may also occur. Water resources systems must adjust to these resulting changes. Sustainability implies robustness, the ability of systems to adjust to changes in system parameters or inputs without requiring excessive expenditures and without losing their ability to meet the demands of current and future generations.

Sustainable water resources systems must adapt to possible changes in climate. Whether temporary or permanent, climate changes affect precipitation amounts, distributions, and temperature. Long-term changes have been observed, for example, in Africa, where some of the large lakes are drying up. Lake Chad and the four large lakes in the African Rift Valley have experienced substantial declines of their water levels over the last 30 to 40 years. Why? While still a matter of scientific debate, many indicators point to a recent rise in global temperature. Past records of the effects of climatic excursions on regions of the earth as well as calculations of climate scenarios by means of general circulation models (GCMs) appear to point toward an increase in local and regional climate changes. If true, these could be accompanied by changes in growth rates of the plant cover and other living organisms. A change in temperature could push the local climate into a different climate zone, having different equilibrium plant covers and ecosystems. These point to the need for flexibility in water resources planning and management.

Should one use the predictions of the GCMs for the planning of a water resources system? This question is difficult to answer. Generally speaking, the main issue of climate change is whether climate in specific regions is permanently or just temporarily changing, and just when and to what extent should action be taken to adapt to that change. Can GCMs really predict the time scale of change? Numerical results published on the prediction of temperature cast doubt on their ability to do so reliably. They, like all models, never represent the entire physical world. They are only abstractions of some aspects of a physical situation, and hence need to be combined with many other aspects to better represent physical reality. Even if one does not go as far as Casti (1989, 1992) who views models as tools for ordering experience rather than as abstractions of reality, one should be reluctant to base long-term decisions on the uncertain predictions of the present generation of models. Yet it is not premature to begin to seriously consider the implications of a Mediterranean climate in Central Europe, or of desertification in Mediterranean Europe.

We know that climate changes have occurred in ancient times. In recent times weather patterns that dominate the climate of Central Europe have apparently changed their frequencies (Bardossy & Caspary, 1990) to produce milder winters and hotter and dryer summers. They also affect other phenomena. For example, they seem to be causing an increase in storm surges coming from the North Sea into the German Bight. Plate (1991) calculated that the probability of floods for the city of Hamburg from storm surges has increased by 45 percent over the last two decades. This change has three causes: (1) human intervention (in this case, the dredging of the lower Elbe River), (2) the sea level has risen (about 4 mm/year), and (3) the frequency of storms over the North Sea has increased. Although storms alone do not cause extremely high tides, the correlation is significant. The sea level rise of the North Sea has lead the low countries bordering it, in particular The Netherlands, to plan and design protection works against extreme tides that might occur in the future.

In such conditions of climatic change uncertainty, the 'precautionary principle' is promoted. As stated in the executive summary to the ASCEND 21 Conference (Jordaan et al., 1993): 'Any disturbance of an inadequately understood system as complex as the Earth system should be avoided.' This implies that all human actions that may contribute to a future climate change should be kept to a minimum. At the same time planners should be alert to possible global changes and take measures commensurate with the time scale of climate change and the time scale of planning, design and construction. A project started today that may take decades for its realization must take climate changes into consideration. Projects with a short time scale must still be designed to be readily adaptable to possible changes in economic, environmental or ecological conditions. It is important, then, to consider that for those developments that will realize adequate benefits over the short term it may be appropriate to plan for their decommissioning should they no longer continue to serve society. For those projects that are expected to provide intergenerational benefits it is necessary to plan for the possibility of excursions from the current climate.

WHAT TO DO?

Given all these issues and challenges with respect to the planning and management of sustainable water resource systems, it is appropriate to ask what can and should be done. No single profession pretends to know enough to answer that question. However with inputs from a multiplicity of professionals and the interested and affected public, resource managers and decision-makers can identify more clearly just what may be done to achieve higher levels of sustainability in specific situations.

Whatever is done to increase the degree of sustainability of our water resources infrastructure will almost certainly involve some costs or require some reduction in the immediate benefits those of us living today could receive. And that is the challenge: deciding what should be done today given what is known as well as what is not, and cannot, be known; in determining how much cost and sacrifice are warranted; and choosing who is going to pay. These issues need to be debated, and this debate should involve everyone having interests in the systems and decisions under discussion.

This challenge – of determining what to do and then getting it done – faces all who choose to assume some responsibility for water resources planning and management. The challenge is one of determining how water and related environmental resources can be developed and managed – managed not only to meet current demands most effectively and efficiently but also to meet the expected future demands. But how can the demands of current populations be satisfied without reducing the options and abilities of future populations to further develop and manage these resource to satisfy their own desires and demands? If that question can be answered, the remaining challenge is one of identifying and implementing programs that satisfy those demands and desires.

Sustainability is an integrating process. It encompasses technology, ecology, and the social and political infrastructure of society. It is probably not a state that may ever be reached completely. But it is one for which we should continually strive. And while it may never be possible with certainty to identify what is sustainable and what is not, it is possible to develop some measures that permit one to compare the performances of alternative systems with respect to sustainability. An attempt to do this will be made in later chapters.

For water resource managers, considerations of sustainability challenge us to develop and use better methods for explicitly considering the possible needs and expectations of future generations along with our own. We must develop and use better methods of identifying development paths that keep more options open for future populations to meet their own, and their descendants', needs and expectations. Finally, we must create better ways of identifying and quantifying the amounts and distribution of benefits and costs (however many ways they might be measured) when considering tradeoffs in resource use and consumption among current and future generations as well as among different populations within a given generation.

SOME THINGS TO REMEMBER

- The continued existence of particular products or technologies or engineering structures, or even of social institutions, is not a necessary condition for sustainable development.
- The non-renewable resources one should consume over space or time for the possibility or promise of something else that is supposedly better than what now exists are difficult to determine. If we maintain too

broad interpretations of sustainable development it will be very difficult to determine progress toward achieving sustainability.

- The consideration of the impacts of what is done now – on present as well as on future generations – is the essence of the term sustainability.

- An explicit consideration of the needs or desires of future generations may require us to think about giving up some of what we could consume and enjoy in this generation.

- Those who are managing natural rsources need to ensure that the public and those who make decisions are aware of the temporal as well as spatial sustainability impacts and tradeoffs associated with those decisions.

- Concern only with the sustainability of larger regions could overlook the unique attributes of particular local economies, environments, ecosystems, resource substitution and human health.

- While probabily most would agree that sustainability involves an explicit focus on at least maintaning (if not increasing) the quality of life of all individuals over time, few would have any idea how it should be done.

- While there is an obligation to begin increasing the sustainability of water and other natural resources today, the practice of planning and management should not wait until we obtain a better understanding.

- Everyone involved in systems development has an obligation to see that they can provide sufficient quantities and qualities, at acceptable prices and reliabilities, without causing environmental deterioriation. If they are not, it is likely that the particular water resource systems are not sustainable.

- There is no doubt but that the perception of sustainability, and thus the motivation for action, depends on an ability to understand the interactions of processes on very different scales.

- A local and regional 'meet the requirements' approach to planning will often fail to consider sustainability.

- Wastewater management methods should be appropriate to the region and its culture and economy.

- While strongly influenced by financing agencies' policies, there is a growing recognition that top-down management plans do not always work well, and that increased attention must be given to user-driven directions.

- It is not necessary that there not be withdrawals from aquifers that have little or no recharge, but rather that some analyses need to be performed to determine just how much should be withdrawn and when.

- The conjunctive use of surface and ground waters can sometimes increase the sustained use of both water systems.

- If those living now must pay for the required remedial measures of any contamination that they produce, they will be less likely to produce it.

- Wherever possible the prevention of pollution is preferable to the reduction or elimination of its consequences.

- Interference by humans on the behavior of a river invariably brings about permanent changes of the river's course. Thus technical river training measures can have both detrimental as well as beneficial effects.

- While surface-water reservoirs can provide many benefits, they can also be the cause of many adverse environmental and social impacts, both upstream and downstream of the dam.

- To increase the sustainability of surface-water reservoirs it is necessary early in the planning stage to identify and carefully study all the beneficial and adverse ecological, economic, environmental and social effects associated with any proposed reservoir project – especially the long-term impacts.

- A guiding principle for the planning of sustainable systems is the concept that a water resource system should interfere as little as possible with the proper functioning of natural life cycles.

- Sustainable system design and operation strongly depends on the capacity building infrastructure available in the region where it is located. Knowledgeable, well-trained and motivated individuals are the backbone of sustainable development.

- An essential aspect in the planning, design and management of sustinable systems is the anticipation of change: changes in the system components due to aging, changes in the demands or desires of society, and changes in the supply of water.

- Whatever is done to increase the degree of sustainability of our water resources infrastructure will almost certainly involve some costs or require some reduction in the immediate benefits those of us living today could receive.

3 Defining sustainability

As stated in Chapter 1, sustainability as defined in the Brundtland Commission's report *Our Common Future* (WCED, 1987) focuses on meeting the needs of both current and future generations. Development is sustainable if,

> . . . it meets the needs of the present without compromising the ability of future generations to meet their own needs.

From the debates that have taken place on sustainable development since that definition was proposed in 1987, one thing is clear: a more specific definition is needed to help those who are engaged in development work to evaluate their efforts with respect to sustainability. Yet in spite of that need, it has been extremely difficult to define just what sustainability is in terms more specific than those suggested by the Brundtland Commission. The previous chapter has outlined some of the issues and challenges water resource systems planners and managers face when attempting to apply this definition.

In this chapter we offer an alternative definition of sustainability specifically applied to water resource system planning and management. This modified definition, we think, can help individuals involved in water resource systems planning and management address some of the issues and challenges mentioned in Chapter 2. It can also help us define some methods for measuring or quantifying the relative sustainability of water resource systems. Such methods are needed if we wish to evaluate alternative plans and policies with respect to their relative sustainability.

While the word *sustainability* can mean different things to different people, it always includes a consideration of the future. But so does 'planning' in general. The Brundtland Commission (WCED, 1987) was concerned about how our actions today will affect, '. . . the ability of future generations to meet their needs.' Just what will those needs be? We today can only guess as to what they may be. We can also argue over whether or not it is appropriate to even attempt to meet needs if and when they over-stress the system designed to meet them. We simply cannot know for certain just how sustainability can be achieved.

Do we enhance the welfare of future generations by preserving or enhancing the current state of our natural environmental resources and ecological systems? Obviously we do, but over what space scales should we do it? How do we allocate over time and space our non-renewable resources, e.g., the water that exists in many deep groundwater aquifers, which are not being replenished by nature? To preserve non-renewable resources now for use in the future, in the interests of sustainability, would imply that those resources should never be consumed as long as there will be future generations. If permanent preservation seems unreasonable, then how much of a non-renewable resource might be consumed, and when? It raises the question, does everything need to be sustained?

If sustainability applies only to human living conditions and standards, as some argue, then perhaps some of today's stock of natural resources should be consumed. The amount consumed today could be used to increase our standard of living, improve our technology, enhance our knowledge, create a greater degree of social stability and harmony and contribute to our culture. All of this might provide future generations with an improved technology and knowledge base that would enable them to further increase their standard of living using even less natural, environmental and ecological resources. Of course, it is impossible to know whether this substitution of natural resources for other capital, intellectual and social resources will happen – or that even if it does happen, whether it would necessarily eventually lead to higher levels of sustainable development and human welfare.

Thus the debate over the definition of sustainability is among those who differ over just what it is that should be sustainable and how to achieve it. Without question, determining who in this debate has the better vision of what should be sustainable and how we can reach a path of sustainable development will continue to challenge us all. But this challenge need not delay our attempts to at least achieve more sustainable water resource systems. In doing so, we may consume some non-renewable resources now and leave some for future generations. To achieve higher levels of sustainability of our renewable water resource systems, we must preserve and enhance their renewing capacity – their capacity to produce the desired amounts and qualities of water, and to support the environment and ecosystems we are all dependent upon. This is certainly a necessary condition if such systems are going to be able to satisfy to the maximum extent possible the 'needs' of future generations, whatever those needs may be.

Do we wish our decisions and actions in this generation to be viewed favorably by future generations? Will future generations find fault with what we decide to do in this generation that may affect what they can do and enjoy in their generation? If we placed all our preferences on future generations, we might define as sustainable those actions that minimize the *regret* of future generations. Clearly we have our current interests and desires too, and indeed there may be tradeoffs between what we wish to do for ourselves now versus what we think future generations might wish us to do now for them. These issues and tradeoffs must be debated and decisions made in the political arena. *There is no scientific theory to help us identify which tradeoffs, if any, are optimum.* Public value judgments must be made about which demands and wants should be satisfied today and what changes should be made to ensure a legacy for the future. Different individuals have different points of view, and it is the combined wisdom of everyone's opinions that will shape what society may consider sustainable.

ALTERNATIVE DEFINITIONS AND PERSPECTIVES

Sustainable development has been defined in many ways and from many perspectives (Serageldin et al., 1993). The United Nations has defined it in terms of the benefits resulting from resource use (UN, 1991). Under this definition an alternative development policy might be considered sustainable if positive net benefits derived from natural resources (including water supplies) can be maintained in the future. There are difficulties, however, in measuring net benefit for some resource uses. For example, how does one evaluate the benefits of wetlands, fisheries, water quality for recreation, ecosystem rehabilitation, preservation, etc.?

Sustainable development has also been defined in terms of financial viability. To be financially sustainable, all costs associated with a development policy should be recovered. The service provided by a water development project must be able to pay for that project. In fact, the revenues should exceed the costs, and thereby provide for the improvement and maintenance of the project. An indication of the financial sustainability of a project paid for by a development bank could be the ability of that project to continue to deliver service or welfare after the initial funding has been spent.

Goodland, Daly & El Serafy (1991) view sustainable development as a relationship between changing human economic systems and larger, but normally slower-changing, ecological systems. In this relationship human life can continue indefinitely and flourish, and the supporting ecosystem and environmental quality can be maintained and improved. Sustainable development, then, is one in which there is an improvement in the quality of life without necessarily causing an increase in the quantity of resources consumed. But the idea of sustainable growth, i.e., the ability to get quantitatively bigger, continually, is an impossibility. Sustainable development, i.e., to improve qualitatively and continually, on the other hand, may be possible.

Falkenmark (1988) defines sustainability based on the role water plays in development. She suggests various conditions for sustainability. Soil permeability and water retention capacity have to be secured to allow rainfall to infiltrate and be used in the production of biomass on a large enough scale for self-sufficiency. Drinkable water has to be available. There has to be enough water to permit general hygiene. Fish and other aquatic biomass have to be preserved and remain edible.

Sustainability and private enterprise

Business and industrial communities are key participants in any program to promote sustainability. They are the primary users (and the primary polluters) of water resources. They may also become the prime developers of water resources when governments are unable to deal with financing and operation of new water projects. For example, in India and China energy demands are so large that government initiatives are stressing direct involvement of private investments in developing thousands of new small hydro projects. The privatization of many formerly public or governmental water resource projects, especially involving hydroelectric power and water supply treatment and distribution, in North and South America and in Europe is well known. Providing services from water resources is increasingly becoming a business enterprise.

Businesses are key creators of wealth, jobs, income and opportunity. Money needed to develop and manage water resources, create jobs, lift people out of poverty, and provide for the demands of growing populations will have to come from economic growth, domestic saving and wise investments at the national and international levels. For developing regions, foreign aid may provide some of this needed capital, but it is always temporary. The remaining, and usually far greater amounts over much longer time periods, must come from the private sector – businesses and industries.

Business leaders can be a force for sustainable development of water and other natural or environmental resources if they are allowed to act as private organizations. That means not being expected to perform public sector chores (such as creating jobs just to reduce unemployment or the number on welfare) and being encouraged through various economic incentives to internalize environmental costs and to produce more with fewer resources, and with less pollution.

Many companies are becoming more 'ecologically conscious' as they respond to a number of pressures and incentives. These include:

- 'green' consumerism
- media emphasis on the environment
- increased willingness of banks to lend to (and of insurance companies to cover) businesses and industries because they will not face big clean-up bills or lawsuits
- internal pressure from employees
- tougher environmental regulations
- new environmental taxes and charges
- and, simply, an increasing emphasis on corporate and personal responsibility to do what is considered right for the common good.

Businesses cannot be expected to achieve sustainable systems, economies and environments by themselves. But more and more of them are willing to be partners with governments and non-governmental organizations, as appropriate, in working together toward achieving this goal.

Business and industry have always fueled development and economic growth. In so doing, they have impacted the environment. They have done this by using environmental resources, by developing and distributing technology and by creating terms and paths of trade. This has lead, and can continue to lead, to an increased quality of life. But an emphasis on short-term profits for shareholders can also lead in the long-term to a degraded environment. This, business and industry must learn to avoid.

Sustainability and social organizations and institutions

Sociologists stress that human patterns of social organization are crucial for devising viable options for achieving sustainable development. Indeed, it is well known that failure to pay sufficient attention to social factors in water resources development processes can seriously jeopardize the effectiveness of development programs and projects (Akowumi, 1994). Sociologists justifiably argue that the creation of stable, culturally appropriate and institutionally enduring patterns of social organization at the grassroots level in support of sustainability objectives is a necessary condition for the achievement of these objectives.

One central social objective is 'quality of life for all'. 'Quality of life' includes the freedom to pursue one's ambitions and express one's abilities, as long as these do not infringe on the freedom and rights of others to pursue the same. 'For all' in this definition means all individuals and groups in the present generation and in all generations to come. The social structures people put in place are usually designed to facilitate the pursuit of this objective rather than as an end in itself. Therefore criteria established such as equal participation and empowerment, are the result of the need to ensure ways of achieving individual quality of life. Thus to sociologists, sustainability would mean an increase, or at least the maintenance, of a quality of life for all.

Sociologists might list the following social objectives for achieving a sustainable water resource system: empowerment, participation, social mobility, social cohesion, cultural identity and institutional development. These objectives might be viewed as equity and poverty reduction by economists, but in the eyes of sociologists they are much more.

Sustainability and natural ecosystems

Ecologists stress preserving the integrity of ecological subsystems as a necessary condition for economic sustainability, and certainly for sustainable water resource systems. Healthy ecosystems are critically important for the overall survival and stability of not only non-degraded water resource systems, but many socio-economic activities as well. Some ecologists argue for the preservation of all ecosystems. A less extreme view aims at maintaining the resilience and dynamic adaptability of natural life-support systems. The units of measure are physical, not monetary, and the prevailing disciplines are biology, geology, chemistry and the natural sciences.

Ecologists might list ecosystem integrity, carrying capacity, biodiversity and reduction of adverse global impacts as ecological objectives for achieving a sustainable system or state. The challenge, however, is how best to measure and predict what happens to the relevant natural systems, or ecosystems, when economic development takes place. Ecologists have been among the leaders in encouraging the integration of ecological considerations into economic analyses. Some ecologists are even advocating a new theory involving the combination of ecology, environment, energy and economics.

Sustainability and economics

Economists and ecologists differ somewhat in their definitions of sustainable development. Much of this debate revolves around the concept of social capital. Ecologists argue that we should focus on the preservation or enhancement of natural resources. Economists argue that we should be looking at the entire mix of resources (the environment, human knowledge, man-made capital, etc.) that comprise what is called social capital. The arguments revolve around the general problem of understanding which elements of social capital our future generations are likely to value the most.

The issue of intergenerational responsibility is central to the concept of sustainability, as is the composition of social capital and scale of human activity. Intergenerational equity requires that each generation manage its resources in ways to ensure that future generations can meet their demands for goods and services, at economic and environmental costs consistent with maintaining or even increasing per capita welfare through time. But how do we know that future generations will value environmental resources as we do? Conditions for sustainability will vary for specific regions, and these differences will increase as the area of the region decreases. Mobility of populations and resources also affect sustainability. It is important, then, to consider the spatial or regional dimensions of sustainability, and the institutional conditions and arrangements that determine the relations among peoples in various regions.

Economists seek activities that will maximize human welfare within the constraints of existing capital and natural stock and technologies. Typical economic objectives for sustainable development might include efficiency, growth and equity (Young, 1992). In an attempt to consider sustainable options, economists are beginning to include in their analyses the additional need to maintain or increase the stock of economic, ecological and socio-cultural assets over time (to ensure the sustainability of income and intragenerational equity) and to provide a safety net to meet basic needs and protect the poor. The challenge, of course, is to determine how to do this. Multiple economic, social, and environmental objectives – such as preserving the dynamic resilience of ecological systems to withstand shocks, promoting public participation or reducing conflicts – are not easily cast in monetary terms. In these cases, economists often rely on other techniques, such as multi-criteria analyses, to facilitate tradeoffs among different objectives.

Sustainability and technology

Technology is created and implemented to solve problems or to make things work better, i.e., to improve people's lives. Engineers create technology. Hence, in their view, whether or not sustainability can be achieved will depend, to a large part, on them. In an article entitled 'Engineering Sustainable Development' Prendergast (1993) writes, '. . . sustainable development is an effort to use technology to help clean up the mess it helped make, and engineers will be central players in its success or failure.' He believes that future technology will use energy and other natural resources more efficiently through conservation measures and switching to renewable sources, waste minimization, much greater recycling and re-use of resources and materials, more comprehensive economic/environmental assessments employing 'life-cycle' analyses and more effective and efficient management of resources.

Civil and environmental engineers like to take credit for having been environmentally conscious for over a century, since the time the first measures were implemented to disinfect drinking water. These engineers have lead the way on the cleanup of rivers and lakes throughout much of the world. They now see themselves in key positions to lead, together with many from other disciplines, the change toward developing a sustainable economy.

Engineers can contribute to sustainable development in two ways: by introducing environmentally beneficial practices within their own organizations, and by becoming involved in and carrying out projects that contribute positively to sustainable development. This is easy to say, but it is not always easy to do. Some engineers will find it difficult, at times, to limit themselves to only those projects contributing to sustainable development, since that may affect their profits and competitiveness. Hence many engineers are discussing among themselves and with their clients various procedures for modifying projects, as necessary, so that such projects can contribute to a more sustainable future. In a few countries this discussion has lead to proposed changes in Codes of Practice (or Conduct), both for engineers and for their clients.

Some engineering firms have undertaken the development of criteria for judging the degree to which a project may contribute to sustainable development. Some of these guidelines are summarized and discussed in Chapter 5 of this monograph.

SUSTAINABILITY AND WATER RESOURCE SYSTEMS

From the previous discussion, it is evident that there is no commonly accepted definition of sustainable development. However, it is clear that sustainability focuses on the long-term improvement of society's welfare. This improvement in welfare over time cannot occur without sustainable water resource systems – systems that can meet, now and on into the future and to the fullest extent possible, society's demands for water and the multiple purposes it serves. These demands will vary from region to region. The demands in each region will include not only the traditional uses of water flows and storage volumes, as applicable, but also the preservation and enhancement of the social, cultural and ecological systems that are dependent on the hydrological regime in the region.

Considering all the definitions and perspectives mentioned above, we propose the following definition:

> *Sustainable water resource systems are those designed and managed to fully contribute to the objectives of society, now and in the future, while maintaining their ecological, environmental, and hydrological integrity.*

This definition differs somewhat from that which requires meeting 'the needs of the present without compromising the ability of future generations to meet their own needs'. At river-basin or regional levels, it may not be possible to meet the 'needs' or demands of even the current generation let alone future generations if those needs or demands are greater than what can be obtained on a continuing basis at acceptable economic, environmental and social costs.

Demand management is every bit as important as supply management.

Furthermore, since it becomes increasingly difficult to estimate what future needs or demands will be as we look farther into the future, it seems evident that our obligation is to ensure that whatever we do today to meet today's needs, we cannot allow renewable water resource systems to be degraded. Assuming future generations will expect at least as much from these same water resource systems as we do, degrading them will reduce their capacity to meet future needs, whatever those needs may be. Degradation prevention applies not only to the ability of water resource systems to provide the desired quantities and qualities of water at acceptable costs and reliabilities, but also to their ability to support the ecological, social and cultural systems necessary for the maintenance and improvement of human welfare.

So, just how can we increase the sustainability of our water resource systems? How can we best meet the needs of today's as well as future generations without system degradation? We believe these decisions must involve everyone, not just the professionals. Decisions taken over time with respect to water resource system design and management must be informed decisions. Providing this information is the job of professionals. As conditions and expectations change, so will our decisions. The sustainable development of water resource systems over time must allow for these changes in priorities and goals.

SOME THINGS TO REMEMBER

- Sustainable water resource systems are those designed and managed to fully contribute to the objectives of society, now and in the future, while maintaining their ecological, environmental, and hydrological integrity.

- At river-basin or regional levels, it may not be possible to meet the 'needs' or demands of even the current generation let alone future generations if those needs or demands are greater than what can be obtained on a continuing basis at acceptable economic, environmental and social costs.

- While the word *sustainability* can mean different things to different people, it always includes a consideration of the future. But so does *planning* in general. We obviously enhance the welfare of future generations by preserving or enhancing the current state of our natural environmental resources and ecological systems, but over what space scales should we do it?

- The challenge over the definition of sustainability often concerns just what it is that should be sustained. But it need not delay our attempts to at least achieve more sustainable water resource systems.

- Clearly we have our current interests and desires, and indeed there may be tradeoffs between what we wish to do for ourselves now and what we think future generations might wish us to do now for them. The issues and tradeoffs must be debated and decisions made in the political arena – there is no scientific theory to help us identify which tradeoffs, if any, are optimum.

- The idea of sustainable growth – the ability to get quantitatively bigger, continually – is an impossibility.

- The money needed to develop and manage water resources, create jobs, lift people out of poverty, and provide for the demands of growing populations will have to come from economic growth, domestic saving and wise investments at the national and international levels.

- Businesses cannot be expected to achieve sustainable systems, economies and environments by themselves. But more and more of them are willing to be partners with governments and non-governmental organizations, as appropriate, in working together toward achieving this goal.

- Healthy ecosystems are critically important for the overall survival and stability of not only non-degraded water resource systems, but many socio-economic activities as well.

- Intergenerational equity requires that each generation manage its recources in ways to ensure that future generations can meet their demands for goods and services, at economic and environmental costs consistent with maintaining or even increasing per capita welfare through time.

- Engineers can contribute to sustainable development in two ways: by introducing environmentally beneficial practices within their own organizations, and by becoming involved in and carrying out projects that contribute positively to sustainable development.

- Decisions must involve everyone, not just the professionals. But it is the job of the professionals to provide the information upon which informed decisions can be made. As conditions and expectations change, so will our decisions.

4   Measuring sustainability

This chapter focuses on ways of measuring relative sustainability – the sustainability of one development or management option in comparison to another. While the focus is on quantifying the concept of sustainability, it is realized that mathematical or technical languages alone are not sufficient to fully measure sustainability. Sustainability involves other aspects that deserve intensive discussion, and requires a willingness to go beyond the scope of what may be quantifiable or measurable. But unless we can measure or describe in precise terms what we are trying to achieve, it becomes rather difficult (if not impossible) to determine how effective we are in achieving or increasing the sustainability of our water resource systems, or even in comparing alternative plans and policies with respect to their sustainability.

EFFICIENCY, SURVIVABILITY AND SUSTAINABILITY

To begin a discussion concerning how one might measure and include sustainability in planning models, it is perhaps useful first to distinguish among several planning objectives that focus on future conditions. These objectives are efficiency, survivability and sustainability (after Pezzey, 1992).

To examine these three objectives, we first need to define them as functions of some variable or variables – the result or outcome of any development or management plan or policy. Hence, assume there is some way we can predict the welfare resulting from whatever decisions we make. Each possible decision that could be made today, denoted by a different value of the index k, will result in a time series of net welfare values, $W(k,y)$, for each period y from now on into the future. Assume there is a minimum level of welfare needed for survival, W_{min}.

A decision, k, is efficient if it maximizes the present value of current and all future net welfare values. Using a discount rate of 'r' per period, an 'efficiency' objective involves a search for the alternative k that will

$$\text{Maximize } \Sigma_y W(k, y)/(1 + r)^y \qquad (4.1)$$

Clearly, as the discount rate, r, increases, the welfare values, $W(k,y)$, obtained now will increasingly contribute more to the present value of total welfare than will the same welfare values obtained sometime in the distant future. In other words, as the discount or interest rate, r, increases, what happens in the future becomes less and less important to those living today. This objective, then, while best satisfying present or current demands, may not always assure a survivable or sustainable future.

Efficiency involves the notion of discounting. There is a time value of assets. It assumes that those who need and could benefit from the use of a given resource today are likely to be willing to pay more for it today than for the promise of having it some time in the future. Yet high discount or interest rates tend to discourage the long-term management of natural resources and the protection of long-term environmental assets. Low discount rates, however, may favor investment in projects which are less likely to survive economically, and which are less likely to invest in environmental protection and the technology needed for efficient resource use and recycling. Thus, the relationship between interest rates, resource conservation and sustainable development is ambiguous (see, for example, Norgaard & Howarth, 1991).

An alternative decision, k, can be considered 'survivable' if in each period, y (on into the future), the net welfare, $W(k,y)$, is no less than the minimum required for survival, W_{min}. Hence if

$$W(k, y) \geq W_{min} \qquad (4.2)$$

then alternative, k, is survivable for all periods, y.

A survivable alternative, and there may be many, is not necessarily efficient (in the normal sense) nor even sustainable.

Next, consider an alternative development path that is sustainable. Here we will consider an alternative as sustainable if it assures that the average (over some time period) welfare of future generations is no less than the corresponding average welfare available to previous generations. Welfare could involve or include opportunities for resource development and use. A 'sustainable' alternative, k, assures that there are no long-term decreases in the level of welfare of future generations. In other words, if

$$W(k, y + 1) \geq W(k, y) \qquad (4.3)$$

the alternative, k, is sustainable for all periods, y.

Equivalently, a sustainable alternative, k, is one that assures a non-negative change in welfare,

$$\mathrm{d}W(k, y)/\mathrm{d}y \geq 0 \qquad (4.4)$$

in each period y. There may be many development paths that meet these sustainability conditions.

The duration of each period y must be such that natural variations in a resource, like water, are averaged out over the period. Results of recent climate change studies suggest these periods may have to be longer than past historical precipitation and streamflow records would suggest due to the increased likelihood of more frequent and longer periods of extremes.

Any constraints on resource use for sustainability reasons must also be judged based on their impacts on economic efficiency and externality conditions (Toman & Crosson, 1991). Before setting them, resource managers should understand how sustainability constraints could affect development paths and policies, especially in regions where substitution among multiple resources is possible but affected by uncertainty and endogenous technical change.

Figure 4.1 illustrates various development paths of net welfare that represent examples of efficient, survivable and/or sustainable development. These development paths are only examples. While not illustrated, it might be possible in some situations to identify a development path that is efficient, survivable and sustainable at the same time. In most cases, however, some tradeoffs are required among these three development objectives.

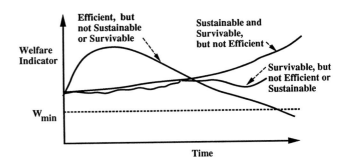

Figure 4.1 Welfare levels, $W(k, t)$, of alternative development paths, k, over time, t. Minimum level considered necessary for survival is shown as W_{min}.

WEIGHTED CRITERIA INDICES

The Delft Hydraulics Laboratory in The Netherlands has proposed a procedure that can be used to measure or quantify the extent to which projects may contribute to sustainable development (Baan, 1994). This procedure consists of responding to a list of criteria, the responses to which may be very subjective. Five main criteria have been identified. Each of the five criteria is subsequently divided into four sub-criteria.

The five main criteria and their respective sub-criteria are:

Socio-economic aspects and impacts on growth, resilience and stability.

- Effects on income distribution
- Effects on cultural heritage
- Feasibility in socio-economic structure

The use of natural and environmental resources including raw materials and discharge of wastes within the carrying capacity of natural systems.

- Raw materials and energy
- Waste discharges (closing material cycles)
- Use of natural resources (water)
- Effects on resilience and vulnerability of nature

Enhancement and conservation of natural and environmental resources, and the improvement of the carrying capacity of natural and environmental resources.

- Water conservation
- Accretion of land or coast
- Improvement and conservation of soil fertility
- Nature development and conservation of natural values

Public health, safety and well-being.

- Effects on public health
- Effects on safety (risks)
- Effects on annoyance/hindrance (smell, dust, noise, crowding)
- Effects on living and working conditions

Flexibility and sustainability of infrastructure works, management opportunities for multifunctional use, and opportunities to adapt to changing circumstances.

- Opportunities for a phased development
- Opportunities for multifunctional use and management and to respond to changing conditions
- Sustainable quality of structures (corrosion, wear)
- Opportunities for rehabilitation of the original situation (autonomous regeneration, active reconstruction and restoration)

Each sub-criterion is given equal weighting. The sum of numerical values given to each sub-criterion is a sustainability index. The higher the index value the greater the contribution of a project to sustainable development. Users of this index must then decide, based on the index value, whether to accept, reject or urge the potential client to modify the project.

WEIGHTED STATISTICAL INDICES ─────────

Alternatively, sustainability indices can be defined as separate or weighted combinations of reliability, resilience and vulnerability measures of various economic, environmental, ecological and social criteria that contribute to sustainability. To use these weighted indices it is first necessary to identify the appropriate economic, environmental, ecological and social criteria to be included in the overall measure or index of relative sustainability. The values of these criteria must be able to be expressed quantitatively or at least linguistically (such as 'poor', 'good' and 'excellent') and be determined from time-series of water resource system variables (such as flow, velocity, water-surface elevation, hydropower production or consumption, etc.).

Criteria that can be expressed in monetary units can be considered economic criteria. This might, for example, include the present value of the economic costs and benefits derived from water resource systems (e.g., from hydropower, irrigation, industry and navigation). Economic criteria usually include distributional as well as efficiency components. Who pays and who benefits is as (if not

more) important as how much the payments or benefits are or will be.

Environmental criteria may include pollutant and other biological and chemical constituent concentrations in the water as well as various hydraulic and geomorphologic descriptors at designated sites of the water resource system. Ecological criteria could include the extent and depth of water in specified wetlands, the diversity of plant and animal species in specified floodplains, and the integrity or continuity of natural ecosystems that can support habitats suitable for various aquatic (including fish) species. Social criteria may include the frequency and severity of floods and droughts that cause hardship or dislocation costs not easily expressed in monetary units, and the security of water supplies for domestic use. They might also include descriptors of recreation opportunities provided by the rivers, lakes and reservoirs, their operation or regulation and the relative quality or attractiveness of the scenery provided for those living next to or using the water resource system.

Once the water resource system is simulated using hydrologic inputs representative of what one believes could occur in the future, the time-series values of these system performance criteria can be derived. These time-series values themselves can be examined in any comparison of alternative water resource system designs and/or operating policies. Alternatively, they can be summarized using the statistical measures of reliability, resilience and vulnerability. The relative sustainability of the system with respect to each of these criteria is higher the greater the reliability and resilience, and the smaller the vulnerability. There are often tradeoffs between these three statistical measures of performance.

To illustrate this procedure, consider any selected criterion, C. Its time series of values from a simulation study are denoted as C_t, where the simulated time periods, t, extend to some future time, T. To define reliability one must identify the ranges of values of this criterion that are considered satisfactory, and thus the ranges of values considered unsatisfactory. Of course these ranges may change within a year and over multiple years. Note that these satisfactory and unsatisfactory ranges of criterion values are subjective. They are based on human judgment or human goals, not scientific theory. In some cases they may be based on well-defined health standards, for example, but most criteria ranges will not have predefined or published standards.

Figure 4.2 illustrates a possible time series plot of all simulated values of C_t along with the designated range of values considered satisfactory. In this example the satisfactory values of C_t are within some upper and lower limits. Values of C_t above the upper limit, UC_t, or below the

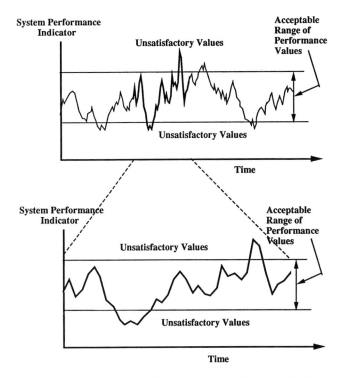

Figure 4.2 Historical or predicted system performance indicator values over time, showing the range of values considered acceptable or satisfactory.

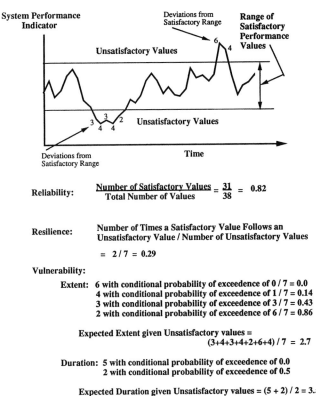

Reliability: $\dfrac{\text{Number of Satisfactory Values}}{\text{Total Number of Values}} = \dfrac{31}{38} = 0.82$

Resilience: Number of Times a Satisfactory Value Follows an Unsatisfactory Value / Number of Unsatisfactory Values

$= 2/7 = 0.29$

Vulnerability:

Extent: 6 with conditional probability of exceedence of $0/7 = 0.0$
4 with conditional probability of exceedence of $1/7 = 0.14$
3 with conditional probability of exceedence of $3/7 = 0.43$
2 with conditional probability of exceedence of $6/7 = 0.86$

Expected Extent given Unsatisfactory values =
$(3+4+3+4+2+6+4)/7 = 2.7$

Duration: 5 with conditional probability of exceedence of 0.0
2 with conditional probability of exceedence of 0.5

Expected Duration given Unsatisfactory values = $(5 + 2)/2 = 3$.

Figure 4.3 Reliability, resilience and vulnerability measures of a predicted system performance indicator.

lower limit, LC_t, are considered unsatisfactory. Each criterion will have its own unique ranges of satisfactory and unsatisfactory values.

Once these data are defined, it is possible to compute associated reliability, resilience and vulnerability statistics, as illustrated in Figure 4.3 and as defined below.

Define the

Reliability of C = Number of satisfactory C_t values /
Total number of simulated periods, T (4.5)

Resilience of C = Number of times a satisfactory C_{t+1}
value follows an unsatisfactory C_t value / Number
of unsatisfactory C_t values (4.6)

Reliability is the probability that any particular C_t value will be within the range of values considered satisfactory. Resilience is an indicator of the speed of recovery from an unsatisfactory condition. It is the probability that a satisfactory C_{t+1} value will follow an unsatisfactory C_t value.

Vulnerability is a statistical measure of the extent or duration of failure, should a failure (i.e., an unsatisfactory value) occur. The extent of a failure is the amount a value C_t exceeds the upper limit, UC_t, of the satisfactory values or the amount that value falls short of the lower limit, LC_t, of

the satisfactory values, whichever is greater. It is the maximum of $[0, LC_t - C_t, C_t - UC_t]$.

The extent of failure of any criterion C can be defined in a number of ways. It can be based on the extent of failure of individual unsatisfactory values or the cumulative extent of failure of a continuous series of unsatisfactory values. In the latter case, each individual extent of failure is added together for the duration of each continuous failure sequence.

Since there can be many values of these individual or cumulative extents of failure for any set of simulations, a histogram or probability distribution of these values can be defined. Thus 'extent-vulnerability' is defined as

Individual Extent-Vulnerability (p) of C
= Maximum extent of individual failure of criterion
C occurring with probability p, or that may be
exceeded with probability $1 - p$ (4.7)

Cumulative Extent-Vulnerability (p) of C
= Maximum extent of cumulative failure of criterion
C occurring with probability p, or that may be
exceeded with probability $1 - p$ (4.8)

Extent-vulnerability can also be defined based on the expected or maximum observed individual or cumulative extent of failure. The conditional expected extent of failure indicator can be defined as

Conditional Expected Extent-Vulnerability of C
= Σ_t individual (or continuous cumulative) extents of failure of C_t / Number of individual (or continuous series of) failure events   (4.9)

The sum of Eq. 4.9 is over all individual or continuous sequences of failure events. Continuous sequences include those of only one period in duration. The unconditional expected extent-vulnerability of C is defined using the above equation except that the denominator is replaced by the total number of simulation time periods, T.

The maximum simulated individual extent of failure can be defined as

Maximum Extent-Vulnerability of
$C = \text{Max}[0, LC_t - C_t, C_t - UC_t]$ (4.10)

For some criteria, for example droughts, the duration as well as the individual and cumulative extents of failure may be important. A histogram or probability distribution of the durations (number of time periods) of failure events can be constructed following any one or multiple simulations. From this histogram or probability distribution, duration-vulnerability measures for each criterion, C, can be defined as

Duration-Vulnerability (p) of C = Maximum duration (number of time periods) of a continuous series of failure events for criterion C occurring with probability p or that may be exceeded with probability $1 - p$ (4.11)
Expected Duration-Vulnerability of C
= Total number of periods t having failures of C_t / Number of continuous series of failure events   (4.12)

The number of continuous sequences of failures for criterion C includes events that last for only one period.

Once these statistical reliability, resilience and vulnerability measures of the time-series values are defined, as appropriate for each economic, ecological, environmental and social indicator or criterion, they can be applied to predicted criteria values over groups of years on into the future. This produces time series of reliability, resilience and vulnerability data for each criteria. If over time these statistical measures are improving, i.e. the reliabilities and resiliencies are increasing and the vulnerabilities are decreasing, the system being studied is getting increasingly sustainable. Often, however, one will find the predicted reliabilities, resiliencies and vulnerabilities of some criteria improving and for other criteria they may be worsening. Or for any given criterion, some of these statistical measures may be improving, and others worsening. This will force one to give relative weights to each measure of each criterion.

One of the better ways to present and examine and compare the values of all these measures for all relevant criteria is through the use of a scorecard. A scorecard is a matrix that presents the values of each of these measures for each of the criteria associated with each alternative system. Table 4.1 illustrates such a scorecard that uses actual values of the criteria that are not assumed to vary over time. Table 4.2 is a scorecard of statistical measures of these criteria that do vary over time. One can see the increasing or decreasing predicted values over each of the five 10-year periods examined in the study on which this scorecard was based.

The scorecard in Table 4.2 shows that some alternatives are predicted to be better with respect to some criteria, and other alternatives are predicted to be better with respect to other criteria. In these situations, which are common, the multi-objective decision-making process will involve making tradeoffs among incommensurate objectives. In some cases this negotiation or decision-making process can be facilitated by attempting to rank each of the alternatives, taking into account each of the criteria. One such approach is through the use of sustainability indices.

To rank each of the alternatives that are non-dominated (i.e., ones that are not inferior to others with respect to all criteria) the information contained in the scorecards can be combined into a single sustainability index. When defining an index using these statistical measures of reliability, resilience and vulnerability, it is convenient to convert each vulnerability measure into a measure that, like reliability and resilience, ranges from 0 to 1 and in which higher values are preferred over lower values. This can be done in two steps. The first involves identifying the largest vulnerability value for each criterion C among all the alternative systems being compared and then dividing each system's vulnerability measure for the criterion by this maximum value. The result is a relative vulnerability measure, ranging from 0 to 1, for each criterion C. One of these relative vulnerability values for each criterion C will equal 1, namely that associated with the system having the largest vulnerability measure.

Relative Vulnerability (C) = Vulnerability (C) / Max Vulnerability (C) among all alternatives   (4.13)

This definition of relative vulnerability will apply to each type of vulnerability identified above, and for any specified level of probability, when applicable.

The second step in converting the vulnerability measure to one that is similar to reliability and resilience, in that higher

Table 4.1. *Example scorecard showing average annual values of various criteria for alternative regional development alternatives. Units of each impact are defined elsewhere.*

Impacts	Alternative policies and components			
	Agricultural irrigation pumps drainage	*Industrial* water storage groundwater use	*Environmental* water recreation treatment tax on water use	*Mixed* water storage treatment canal transport
Annual investment costs	300	400	700	700
Annual economic benefits	1200	700	100	1000
Agricultural production	800	150	50	600
Drinking water cost	1.40	0.90	1.20	1.10
Pollution index at site A	150	220	30	70
Power production	200	1200	50	800
Fisheries production	70	20	80	40
Flood production %	99	98	96	99

Best (□) Worst (▢)

values are preferred over lower values, is to subtract each relative vulnerability measure from 1.

Once this scaling and conversion have been performed, each statistical measure ranges from 0 to 1 and higher values are preferred over lower values. They can now be combined into a single index for each criterion, C. One way of doing this is to form the product of all these statistical measures.

$$\text{Sustainability } (C) = [\text{Reliability } (C)] \, [\text{Resilience } (C)]$$
$$[\Pi_v \{1 - \text{Relative Vulnerability } v(C)\}]$$
(4.14)

where Relative Vulnerability $v(C)$ is the vth type of relative vulnerability measure being considered for criterion C. The use of multiplication rather than addition in the above index gives added weight to the statistical measure having the lowest value. For example, if any of these measures are 0, it is unlikely any of the other measures are very relevant. A high value of the index can result only if all statistical measures have high values.

The resulting product, the Sustainability(C) index, ranges from 0, for its lowest and worst possible value, to 1, at its highest and best possible value. This sustainability index applies to each criterion, C, for any constant level of prob-

ability, p, and can be calculated for each alternative system or decision being considered.

To obtain a combined weighted relative sustainability index that considers all criteria, relative weights, W_C, ranging from 0 to 1 and summing to 1, can be defined to reflect the relative importance of each criterion. These relative weights may indeed be dependent on the values of each Sustainability (C) index. Once defined, the relative sustainability of each alternative system being compared is

$$\text{Relative Sustainability } = \Sigma_C W_C \text{ Sustainability } (C) \quad (4.15)$$

Since each sustainability index and each relative weight W_C ranges from 0 to 1, and the relative weights sum to 1, these relative sustainability indices will also range from 0 to 1. The alternative having the highest value can be considered the most sustainable *with respect to the criteria considered, the values of each criterion that are considered satisfactory and the relative weights.* Every one of these assumptions involves the subjective judgments of those participating in the evaluation process.

Note what has been done here. A particular proposed or actual water resource system design and management policy is simulated over time. The simulated data include assump-

Table 4.2. *Example scorecard showing average annual values and trends in reliability, resilience and relative invulnerability of various criteria values for alternative regional development alternatives. Units of all criteria are defined elsewhere. Relative weights of sustainability index were assumed to be equal and sum to 1.*

Impacts	Alternative policies and components			
	Agricultural	*Industrial*	*Environmental*	*Mixed*
	irrigation pumps drainage	water storage groundwater use	water recreation treatment tax on water use	water storage treatment canal transport
Annual investment costs	300	400	700	700
Reliability trends	0.89–0.93	0.85–0.89	0.82–0.84	0.88–0.89
Resilience trends	0.77–0.76	0.78–0.80	0.67–0.75	0.74–0.78
1−Rel. vulnerability				
Extent trends	0.84–0.91	0.75–0.77	0.66–0.72	0.78–0.83
Duration trends	0.91–0.96	0.92–0.90	0.89–0.87	0.90–0.92
Overall Product	0.52–0.62	0.46–0.49	0.32–0.39	0.46–0.53
Annual economic benefits	1200	700	100	1000
Reliability trends	0.83–0.88	0.82–0.85	0.82–0.88	0.78–0.89
Resilience trends	0.71–0.81	0.76–0.83	0.77–0.79	0.84–0.88
1−Rel. vulnerability				
Extent trends	0.94–0.91	0.79–0.76	0.76–0.77	0.88–0.89
Duration trends	0.91–0.86	0.90–0.93	0.85–0.88	0.93–0.96
Overall Product	0.50–0.56	0.44–0.50	0.41–0.47	0.54–0.67
Agricultural production	800	150	50	600
Reliability trends	0.92–0.94	0.65–0.60	0.72–0.77	0.78–0.79
Resilience trends	0.71–0.75	0.72–0.76	0.85–0.90	0.76–0.77
1−Rel. vulnerability				
Extent trends	0.88–0.93	0.62–0.67	0.86–0.92	0.88–0.84
Duration trends	0.93–0.86	0.71–0.75	0.89–0.85	0.90–0.95
Overall Product	0.53–0.56	0.21–0.23	0.47–0.54	0.47–0.49
Drinking water cost	1.40	0.90	1.20	1.10
Reliability trends	0.83–0.85	0.88–0.89	0.82–0.86	0.85–0.91
Resilience trends	0.81–0.80	0.88–0.85	0.77–0.78	0.78–0.88
1−Rel. vulnerability				
Extent trends	0.84–0.93	0.85–0.87	0.86–0.92	0.88–0.93
Duration trends	0.92–0.95	0.82–0.90	0.89–0.88	0.90–0.91
Overall Product	0.52–0.60	0.54–0.59	0.48–0.54	0.53–0.68

Table 4.2. *Continued*

| Impacts | Alternative policies and components | | | |
| | *Agricultural* | *Industrial* | *Environmental* | *Mixed* |
	irrigation pumps drainage	water storage groundwater use	water recreation treatment tax on water use	water storage treatment canal transport
Pollution index at site A	150	220	30	70
Reliability trends	0.79–0.83	0.75–0.79	0.92–0.94	0.78–0.83
Resilience trends	0.73–0.77	0.76–0.80	0.97–0.95	0.90–0.92
1−Rel. vulnerability				
Extent trends	0.74–0.81	0.71–0.73	0.86–0.92	0.88–0.89
Duration trends	0.81–0.86	0.82–0.90	0.89–0.94	0.84–0.88
Overall Product	0.35–0.45	0.33–0.42	0.68–0.77	0.52–0.60
Power production	200	1200	50	800
Reliability trends	0.78–0.80	0.77–0.76	0.67–0.75	0.74–0.83
Resilience trends	0.85–0.89	0.89–0.93	0.82–0.84	0.88–0.92
1−Rel. vulnerability				
Extent trends	0.92–0.90	0.91–0.96	0.89–0.87	0.90–0.89
Duration trends	0.75–0.77	0.84–0.91	0.66–0.72	0.78–0.78
Overall Product	0.46–0.49	0.52–0.62	0.32–0.39	0.46–0.53
Fisheries production	70	20	80	40
Reliability trends	0.82–0.84	0.78–0.83	0.84–0.91	0.92–0.90
Resilience trends	0.68–0.75	0.90–0.92	0.91–0.96	0.75–0.77
1−Rel. vulnerability				
Extent trends	0.65–0.72	0.88–0.89	0.89–0.93	0.78–0.80
Duration trends	0.89–0.87	0.74–0.78	0.77–0.76	0.85–0.89
Overall Product	0.32–0.39	0.46–0.53	0.52–0.62	0.46–0.49
Flood Protection %	99	98	96	99
Reliability trends	0.89–0.98	0.95–0.99	0.90–0.92	0.89–0.87
Resilience trends	0.97–0.96	0.78–0.80	0.74–0.83	0.67–0.72
1−Rel. vulnerability				
Extent trends	0.94–0.91	0.85–0.87	0.78–0.78	0.66–0.75
Duration trends	0.91–0.92	0.92–0.95	0.88–0.89	0.82–0.84
Overall Product	0.71–0.81	0.58–0.65	0.46–0.53	0.32–0.39
Overall sustainability index	0.55–0.56	0.44–0.50	0.46–0.53	0.47–0.55

Best		Worst

tions regarding system design, operation, and hydrologic and other inputs and demands that represent a scenario representative of what could occur in the future. Incorporated into that simulation are the variables individual stakeholders consider important and relevant to sustainability. These are called criteria. These criteria could include physical, economic, environmental, ecological and social variables. A time series of each of these criterion values at different locations are produced from the simulation. These time series are divided into sub-periods and statistical measures of reliability, resilience and vulnerability are used to summarize each sub-period's time series of each criterion. The results can be presented on scorecards, or can be combined into a sustainability index for each of the simulated criterion variables considered important with respect to system sustainability. These individual indices for succeeding time periods can be compared to judge whether sustainability is increasing or decreasing over time.

Alternatively the sustainability index value of each criterion can be combined, using relative weights, into an overall sustainability index, and this combined measure can be used to judge whether or not sustainability is increasing over time.

Now, consider some complications. These criterion values are in all likelihood spatially dependent. It is very likely that these relative sustainability indices will vary depending on the site where the time-series values are observed and computed. If so, these relative sustainability indices can be computed for various sites for each alternative water resource system being evaluated and compared. Each of these site-specific relative sustainability indices can then be considered using scorecards and other multi-objective analyses methods. Alternatively, one could develop an overall system indicator of reliability, resilience and vulnerability through some averaging scheme.

Another complication can occur when dividing the time-series data into sub-periods over which these statistical measures are to be defined. This partitioning must be done so as to average out the natural variation in hydrologic systems but not those variations caused by human activities. The main purpose of this exercise is to identify trends that indicate a worsening situation or an improving situation over time. Clearly the values of these relative sustainability indices will be dependent on the durations of the sub-periods.

It is important to remember that these relative sustainability indices are all based on subjective assumptions concerning future hydrology, costs, benefits, technology, ecological responses and the like. They are also based on subjectively determined ranges of satisfactory or unsatisfac-

tory values, and, if used, on subjectively determined relative weights. Nevertheless, if the criteria used are comprehensive and identify the concerns and goals of everyone now and, we hope, on into the future, as best we can guess, relative sustainability indices can be used to help identify preferred design or policy alternatives. They can be used along with other criteria, as applicable, in a multi-objective analysis.

> There is no guarantee that analyses such as these, performed by different groups and/or at different times, will end up with the same conclusions.

Once again, change is with us, and different groups within and over different generations will have different views of just what is sustainable. Using methods such as the ones just proposed, however, forces all of us to look into the future as best we can and to evaluate the multiple physical, economic, environmental, ecological and social impacts of what we decide to do now on individuals living in future generations. It also ensures that we have that information in some summarized and comparable form available for everyone involved in the decision-making process.

REVERSIBILITY, ROBUSTNESS AND SUSTAINABILITY

Given the difficulty of predicting, with any reasonable degree of certainty, just what will happen some 50, 100, 150, etc. years from now, and just what those living then will have as their goals and needs, many suggest that we shouldn't even attempt such predictions. They argue that the best we can do today to enhance sustainability is to maintain system reversibility and robustness. Reversibility implies keeping design and management options open or available for future generations. Robustness refers to the system's ability to adapt, at minimal costs, to varying, often unforeseen, conditions in the future. System robustness has also been called design flexibility in some of the economic literature. Robust, or flexible, systems may not be the most cost-effective for the expected future conditions, but rather they are designed to be near cost-effective for a wide range of possible future conditions. Hashimoto et al, (1982) are among those who have applied this economic concept to the evaluation of alternative water resource system designs.

Irreversibility has been measured in terms of entropy – the degree to which processes reduce options future generations may want to consider. Among those who have examined the use of entropy as a measure of irreversibility and sustainability are McMahon & Mrozek, (1996); Meadows *et al.* (1992); Rifkin (1989); and Ruth (1993) to cite a few. Nachtnebel

(1996) applied the concept of entropy to the irreversibility of groundwater resource systems. In all of these analyses, energy is the principle parameter of concern. For many, it replaces the need for time-series simulations and analyses. It also replaces the need for human preference estimations. Both involve large uncertainties. Clearly, more research on the use of entropy as a sustainability criterion will be forthcoming in the future.

CONCLUSION

This chapter has outlined a number of ways sustainability criteria might be quantified and used to compare alternative plans, designs and policies with respect to their relative sustainability. No doubt there are many others that can be defined and should be used. The remaining sections of this monograph identify and address in more detail some of these sustainability criteria, their relationships to various water resources variables or characteristics, and their relative importance in the planning and managing of water resource systems in the foreseeable future.

SOME THINGS TO REMEMBER

- When considering efficiency, as the discount or interest rate, r, increases, what happens in the future becomes less and less important to those living today. This objective, then, while best satisfying present or current demands, may not always assure a survivable or sustainable future.

- Sustainability indices can be defined as separate or weighted combinations of reliability, resilience and vulnerability measures of various economic, environmental, ecological and social criteria that contribute to sustainability.

- Satisfactory and unsatisfactory ranges of criterion values are subjective. They are based on human judgment or human goals, not scientific theory. In some cases they may be based on well-defined health standards, for example, but most criteria ranges will not have predefined or published standards.

- For some criteria, for example droughts, the duration as well as the individual and cumulative extents of failure may be important. One of the better ways to present and examine and compare the values of all these measures for all relevant criteria is through the use of a scorecard.

- Partitioning of time series must be done so as to average out the natural variation in hydrologic systems, but not those variations caused by human activities. The main purpose of this exercise is to identify trends that indicate a worsening situation or an improving situation over time.

- It is important to know that there is no guarantee that analyses concerning criteria, performed by different groups and/or at different times, will end up with the same conclusions.

- Some would argue that given the difficulty of predicting, with any reasonable degree of certainty, just what will happen some 50, 100, 150, etc. years from now and just what those living then will have as their goals and needs, we shouldn't even try such predictions. It is suggested that the best we can do to enhance sustainability is to maintain system reversibility and robustness.

5 Sustainability guidelines and case studies

THE FOCUS OF GUIDELINES

This chapter reviews some guidelines that have been designed by and for professionals involved in water and environmental resource development and management. These guidelines have been obtained from documents published by various institutions and government agencies throughout much of the world. The guidelines listed below have been developed in Asia, Australia, Europe, and North America. They focus on effectively managing and using our natural and human resources and at the same time preserving the quality of the environment and its ecosystems. These guidelines were developed to help ensure that our water resource systems are

> ... designed and managed to fully contribute to the objectives of society, now and in the future, while maintaining their ecological, environmental, and hydrological integrity.

We have combined the guidelines we have found into six separate topics:

1. The design, management and operation of the physical infrastructure
2. The environment and ecosystems
3. Economics and finance
4. Institutions and society
5. Health and human welfare
6. Planning and technology

The guidelines in each of these six areas serve as goals for water resources engineers, planners and other professionals who are responsible for, and who arguably should take the lead in, increasing the sustainability of our water resource systems.

SAMPLE GUIDELINES FOR WATER RESOURCE SYSTEMS ENGINEERS AND PLANNERS

Design, management and operation of physical infrastructure

Those responsible for water resource systems planning, design and management should:

- Design and manage systems to be effective, efficient and robust in all respects – balancing changes in demands and supplies over time and space.
- Ensure that systems can adjust to changing land uses without necessitating excessive construction expenses.
- Ensure that implementation, operation, maintenance and management of water resources projects are undertaken by those most knowledgeable of the needs of those for which the systems serve, and that they have opportunities to continually improve their skills and knowledge.
- Ensure that human actions and activities do not impair the long-term health and resilience of freshwater stocks and flows.
- Ensure that systems are resilient to failure, i.e., all components of them can be replaced or repaired without undue disruption of services.
- Design and operate reservoirs and manage upstream watersheds so as not to reduce significantly their active storage capacities.
- Ensure that provisions are in place and that personnel are trained to cope with natural and/or man-caused disasters and changes in demand, supply, land use and climate.

- Ensure that management continuously monitors the performance of the systems and assesses and improves the total multi-criteria system performance, adapting to changing conditions and goals as appropriate.
- Ensure that systems conserve renewable water resources, and limit their use to their sustainable yield.
- Ensure that systems conserve and effectively use non-renewable water resources and employ recycling when appropriate.
- Use demand management in conjunction with supply management.

Environment and ecosystems

Those responsible for water resource systems planning, design and management should, with respect to the environment and ecosystems:

- Ensure that water quality is considered along with water quantity when designing and operating water resource systems.
- Ensure that water quality is maintained to meet certain minimum standards that may vary over time and space.
- Ensure that there are no negative long-term irreversible or cumulative adverse effects on the environment or on its ecosystems.
- Guarantee sufficient water regimes to maintain and restore, if applicable, the health of aquatic and floodplain ecosystems.
- Evaluate and monitor the beneficial and adverse environmental impacts and take actions to alleviate the adverse ones.
- Ensure that systems interfere as little as possible with the natural environment and that any interferences are such that they can be absorbed by the environment without adverse consequences.
- Protect areas used for the collection of surface and ground water against pollution and inappropriate land use.
- Protect and restore the land and natural ecosystems associated with the water resource systems.
- Eliminate non-natural soil erosion in catchment areas.
- Protect and enhance the aesthetic environment.
- Maintain interdependence and diversity of our natural ecosystems that form the very basis of our continued existence.
- Acknowledge and respect the finite capacity of the environment to assimilate changes due to human activities.

- Incorporate environmental objectives, conservation and energy efficiency into the design and operation of engineering facilities, to prevent or minimize any adverse environmental effects.
- Take any actions required to restore and sustain the natural environment and its ecosystems as needed in specific situations.
- Ensure that projects include post-development monitoring of environmental changes and adjust operations as a result of that monitoring.
- Make sure that any errors are made on the safe side with respect to environmental consequences, since the response of biological systems to human activities is frequently difficult to predict.

Economics and finance

Those responsible for water resource systems planning, design and management should, with respect to economics and finance:

- Fully consider all direct and indirect environmental costs over the full life cycles of the systems' projects.
- Recover all costs of all resource development and management projects throughout their life cycles in an equitable and efficient way.
- Make sure that society supports and is willing to pay for the services provided by the water systems.
- Ensure that sufficient finances are available to continuously operate and monitor the performance of water resource projects.
- Reduce systems operating costs, including that for energy, as much as possible.
- Distribute all system costs and benefits equitably within the user community.
- Include costs and benefits related to environmental quality in economic evaluations of engineering activities.

Institutions and society

Those responsible for water resource systems planning, design and management should, with respect to institutions and society:

- Establish effective procedures to manage conflicts over water management and use.
- Implement fully democratic and participatory water planning and decision-making processes, involving all stakeholders in the planning, execution and management of the systems as much as possible.

- Ensure that professionals and the public have a broad understanding of the political, economic, scientific and social issues that will impact their involvement and interaction with respect to water resources planning and management.
- Create the political will and provide the leadership to plan, construct and operate the water resources systems in what is considered a sustainable way.
- Minimize the potential for future international conflicts as much as possible.
- Implement institutional procedures to aid in reaching acceptable solutions to conflicts among stakeholders without unnecessary costs and litigation.
- Ensure that responsible institutions have the capacity to plan, manage, monitor and adapt to changing situations.
- Guarantee the rights of, and encourage, the community to be involved in project formulation, development and operation.

Health and human welfare

Those responsible for water resource systems planning, design and management should, with respect to health and human welfare:

- Guarantee a minimum water supply to all humans to maintain human health.
- Minimize all adverse social impacts caused by dislocations of people and stress during a system failure (such as water shortage or a flood or toxic contamination) and preserve and protect society's cultural heritage.
- Recognize and maintain the worth of architectural, engineering, historic, cultural, archaeological and scientific sites and their existing structures and of the living systems within them.
- Provide for, protect and preserve all high-quality, unique and rare natural resources and systems in system planning, design and operation.
- Evaluate and consider the consequences of all plans, policies and actions – direct or indirect, immediate or long-term – upon social security, human health and equity.

Planning and technology

Those responsible for water resource systems planning, design and management should, with respect to planning and technology:

- Recognize that planning is multi-disciplinary by nature, and includes evaluation of all relevant options, including non-structural solutions and consideration of long-term effects of options (with preference being given to achieving long-term over short-term benefits), and incorporation of conservation objectives into design criteria.
- Collect and make available to all interested individuals all data on water resource availability, use and quality.
- Continuously improve the data on which a water resources system depends and upgrade operating rules to reflect changes in both the data base and in the demands on the system.
- Select non-structural solutions rather than structural ones, whenever possible.
- Maintain options for future uses of resources.
- Recognize individual limitations in assessing sustainability criteria and the need to involve all stakeholders in the planning and management of such systems so that projects can be made compatible with local living conditions and local environments.
- Consider in the planning and design stage the potential future changes in the use of the systems that might be needed to meet possible changing societal demands.
- Ensure that plans are available, and periodically updated, for managing risk, and should failure occur, that individuals are prepared to handle the consequences.
- Promote and implement the development of alternatives to the use of non-renewable water resources and the wise use of non-renewable resources through waste minimization and recycling.
- Ensure that systems achieve their beneficial objectives with the lowest possible consumption of raw materials and energy, both during and after construction.
- Consider in the planning and decision-making processes the long-term planning gains along with the more usual (and biased) short-term planning returns.
- Include the quality of life objectives, both for current and future generations in all planning and decision-making processes, including the knowledge that flora and fauna not only have the right to exist, but are vital to the health and well-being of humans.

CASE STUDIES

Given these broad guidelines that have been established by governmental and professional organizations in various parts of the world, it is reasonable to ask if, and how well, these guidelines have been, and are being, followed. The remainder of this chapter contains a number of brief descriptions of real water resources development projects. Emphasis is given to actions that have been taken to better satisfy society's immediate objectives and the extent to which these actions may, or may not, have increased the sustainability of those systems.

We cannot evaluate or judge the merit of the decisions made in each of these case studies. We present them as examples of what has been done in support of, or in opposition to, some of these sustainability guidelines. There are reasons for each decision taken, many of which we certainly do not and cannot know. Nevertheless these brief descriptions can be examined using these sustainability guidelines, as applicable, as a filter. This permits some judgment, based on what is written, of how those who have planned, designed, and managed water resource systems in this generation may be judged by future generations with respect to the sustainability of their decisions.

These brief descriptions of various water resources projects are not ex-post evaluations of these projects. Not all of the original objectives of the projects are identified, analyzed or evaluated as to how well or how poorly they have been met over time. And as time passes, objectives change.

Today many of these 'past projects' are looked at through 'new glasses', i.e., with respect to the sustainability guidelines presented in the previous section. When many past decisions were made, these 'glasses' were not available; Sustainability was not being considered as an important objective. It is left to the reader to conclude, based on the partial information presented, just how well or how poorly the projects presented below were planned, managed and operated with respect to today's sustainability criteria. Will future generations welcome these actions? Was the technology used and implemented appropriate? Would the decisions made in the past be made today or tomorrow, as best we can guess? If not, how would they differ? How might the process of making decisions differ?

The Aral Sea (south central Asia)

The Aral Sea and its basin, shown in Figure 5.1, in south central Asia, is drying up. Because of the large diversions of water (particularly for irrigation) that would normally have flowed into the sea during the past thirty years, the volume of the sea has decreased to a third of its original 1960 volume. The sea's surface has decreased, the water in the sea and in the surrounding aquifers has become increasingly saline, and

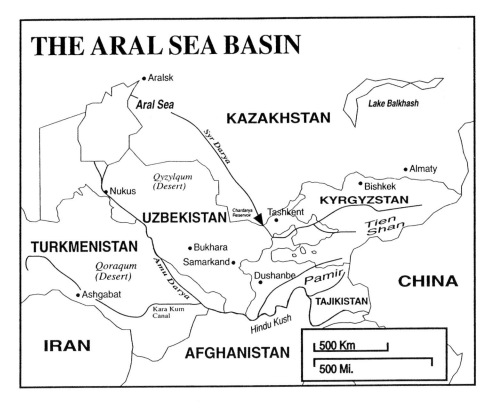

Figure 5.1 The Aral Sea and its basin in south central Asia.

the water supplies and health of almost 40 million people in the Aral Sea basin are threatened. Large areas of salty flatlands have been exposed as the sea has receded, and salt from these areas is being blown across the plains onto neighboring croplands and pastures, causing ecological changes.

The frost-free period in the delta of the Amu Darya River, which feeds the Aral Sea, has fallen to less than 180 days – below the minimum required for growing cotton, the region's main cash crop. The changes in the sea have effectively shut down a substantial fishing industry, and the variety of fauna in the region has declined considerably. If current trends continue unchecked, the sea is predicted to eventually shrink to a saline lake about one-sixth of its 1960 size.

This ecological disaster, as it is commonly called, is the consequence of excessive abstraction of water for irrigation purposes from the rivers that feed the Aral Sea. Total river runoff into the sea has decreased from an average 55 km^3 per year in the 1950s to only a few km^3 per year in the early 1980s. The irrigation schemes have been a mixed blessing for the populations of the Central Asian republics (Kazakhstan, Kyrgyzstan, Tajikistan, Turkmenistan and Uzbekistan) which they serve. The diversion of water has undoubtedly provided livelihoods for the region's farmers, but at considerable environmental and long-term economic cost. The agricultural productivity of soils has been decreased due to accumulation of salt, over-watering has turned pasture land into bogs, water supplies have become increasingly polluted by pesticides and fertilizer residues, and the deteriorating quality of drinking water and sanitation is adversely affecting human health.

While it is easy to see how the problem of the Aral Sea might have been avoided, possible solutions for its reclamation are not obvious. A combination of better technical management and appropriate incentives is clearly essential. For example, charging for water or allocating it to the most valuable uses could prompt shifts in cropping patterns and make more water available to industry and households.

But the changes needed are vast, and there are not many options. The inhabitants of Central Asian Republics (now countries) are for the most part very poor. Their incomes are 65 percent of the average in the former USSR. The countries themselves are in various states of transition from centrally planned to market economies. In the past, transfers from the central government exceeded 20 percent of national income in Kyrgyzstan and Tajikistan and 12 percent in Uzbekistan. These transfers are no longer available. The regional population of 35 million is growing rapidly in spite of high infant mortality. The states have become dependent on a specialized but unsustainable pattern of agriculture. Irrigated

production of cotton, grapes, fruit and vegetables accounts for the bulk of export earnings.

Any rapid reduction in the use of irrigation water will reduce living standards further unless these economies receive assistance to help them diversify away from irrigated agriculture. Meanwhile, salination and dust storms erode the existing land under irrigation.

This is one of the starkest examples of the need to combine development with sound environmental policy (da Cunha, 1994). Clearly in the region of the Aral Sea the answer regarding sustainability is obvious. Is it also obvious when considered from the point of view of the former USSR as a whole? In any region, should some of its land and water resources be degraded in an attempt to increase the probability that the remaining region will become more sustainable? In this case such decisions, if made conscientiously, didn't work. It remains for those now living in the Aral Sea region to begin the process of recovery in a far more complex institutional environment than existed during the development of irrigation in this river basin. This will likely take considerable time, even with substantial outside technical assistance and funding.

This case is one of the starkest examples of the need to consider the environmental consequences of a project and management policy. The negative effects on the Aral Sea itself, and the consequences with respect to health, environment, and economic aspects most probably outweigh the benefits, which accrued to the region's population over the years from the vast irrigation schemes. Had these consequences been anticipated and taken into account at the time the scheme was planned, it is possible that the plans would have been different. But they may not have.

It is possible that the economic and social capital accumulated by the project, i.e. the project benefits to future generations has been far greater than the losses. Even if overall benefits have not exceeded the obvious costs, perhaps the planners of the irrigation projects thought they would. It is also possible that without this large irrigation project many people may not have survived economically, even in the short run. People need to eat and feel useful in society before they concern themselves with environmental, ecological, and other long-term sustainability issues. This raises the equity issues associated with sustainability, equity over both time and space. How can this be achieved? Just what is equitable?

The Ogallala Aquifer (USA)

The Ogallala Aquifer is situated in the mid-western USA (Figure 5.2). Its some 174,000 square mile area underlies

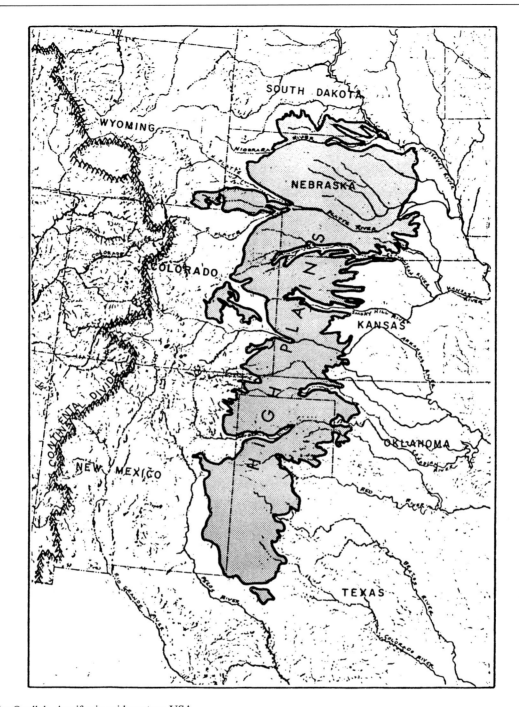

Figure 5.2 The Ogallala Aquifer in mid-western USA.

parts of the states of South Dakota, Wyoming, Nebraska, Colorado, Kansas, Oklahoma, New Mexico and Texas. The aquifer is part of the nation's largest underground water storage system.

The Ogallala Aquifer, also known as the 'High Plains Aquifer,' is the buried erosional remnant of the Rocky Mountains. The Ogallala is not an underground lake nor an underground river as envisioned by many, but a gigantic sponge holding enough water to fill Lake Huron. The aquifer ranges in thickness from less than a foot to 1,300 feet and averages about 200 feet in most areas. The greatest volume of water in the Ogallala underlies the state of Nebraska where the aquifer actually bubbles to the surface in places.

The region has highly productive soils, a relatively level terrain and a temperate climate. Although it underlies only about 6 percent of the land in the USA, it includes almost 13 million hectares of cropland and produces over 15 percent of the total value of wheat, corn, sorghum and cotton and about 38 percent of the total value of the livestock produced in the country.

The surface area on top of the Ogallala Aquifer is dry and very windy. Normal annual precipitation is only about 46 cm/year. Hence, over 40 percent of the cropland in the area is irrigated. The prevailing westerly winds induce high evaporation rates and wind-drift water losses for conventional sprinkler irrigation systems.

Over 90 percent of the irrigation water used in the area is pumped from the Ogallala Aquifer. The aquifer depends almost entirely on the on-site precipitation for recharge. In the southern portion of the aquifer the recharge rate is estimated to average only about 1 cm/year, a mere 20 percent of what would be needed to replace the water currently being withdrawn for irrigation. Nevertheless, the aquifer, which stores approximately 3.6×10^9 m^3, is a major source of high quality water. At one time there was enough water in the aquifer to satisfy US irrigation demands for decades. However excessive and inefficient irrigation has reduced aquifer storage substantially.

Since the end of World War II, modern pumping technologies have made it possible to use increasing amounts of water from this aquifer for irrigation. In 1950, the total irrigated area was about 1.4 million hectares. During the 1950s, after years of intensive pumping for irrigation, the yearly drop in the water table began to be noticed. In 1980, the irrigated area had reached close to 5.7 million hectares and an estimated 23% of the underground water was gone. In west Texas the depletion was about 40%. The water table has dropped more than 15 m over 25 percent of the area since 1940.

Presently, most of the water pumped from the Ogallala is used to irrigate crops, much of which is used to feed livestock. It is estimated that from the 1940s to 1980 the aquifer declined an average of ten feet with some areas in Texas declining nearly 100 feet. Southwestern Kansas shows some of the most severe long-term drawdowns. In parts of that state, wells show an average decline of 150 feet since large-scale irrigation began in the 1950s. In places, more than half of the water originally in place in the Ogallala has been pumped away. Fewer declines were measured in west-central Kansas wells in 1994, and water levels were stable or increased from one to five feet in many areas. Long-term declines of more than 50 feet, however, are common in these same areas.

However, during the 1980s, the aquifer declined only another foot due to improved irrigation practices and new

technologies. As a result of this depletion, the last decade has seen a 20 percent decline in areas under irrigation in the region. Several factors, including increased pumping costs caused by declining groundwater elevations, rising fuel costs and relatively low commodity prices are responsible for the decline. This in turn has decreased the aquifer's groundwater depletion rate. Improved conservation practices have also helped reduce the rate of depletion. These practices include advanced sprinkler systems, putting some cropland in grass to allow the aquifer to recharge itself naturally and using soil moisture monitoring sensors to determine optimum groundwater allocations for irrigation.

Mining Water from the Sahara in Libya

For centuries the vast deserts of southern Libya formed a barrier crossed only by caravan trade routes which followed established tracks from oasis to oasis. Since 1953, these vast and largely unknown areas have been progressively investigated in the search for new oil fields. This lead not only to the discovery of large oil reserves but also to great quantities of fresh water.

During the ice ages in northern Europe, the climate of North Africa became temperate and there was considerable rainfall. The excess rainfall infiltrated into the ground and was trapped in the porous rocks between impermeable layers, forming reservoirs of fresh ground water. The majority of this fresh water is between 14,000 and 38,000 years old.

Four major underground basins have been located in Libya (Figure 5.3). The Kufra basin of 20,000 km^3 storage capacity (at a depth of over 2,000 m, with water of excellent quality), the Sirt basin (10,000 km^3, at 600 m), the Murzuk basin (4,800 km^3, at 800 m) and the Hamadah and Jufrah basins that are underlain by the Paleozoic aquifer.

The expanding economy and growing population along the fertile coastal strip of the Socialist People's Libyan Arab Jamahiriya is creating an increasing demand for water for irrigation, for industry and for domestic and municipal use. At the same time, the traditional groundwater resources are becoming increasingly at risk through intensive use, which is resulting in saline intrusion of the coastal aquifer. This phenomenon would, if unchecked, turn agricultural lands into infertile sabkha.

Extraction of the water known to lie below the desert had been contemplated for many years. In 1974, Libya took the first steps toward exploitation of these water resources when studies were commenced to develop and implement the Great Man-Made River Project. This project will deliver large quantities of water from deep in the desert, over long

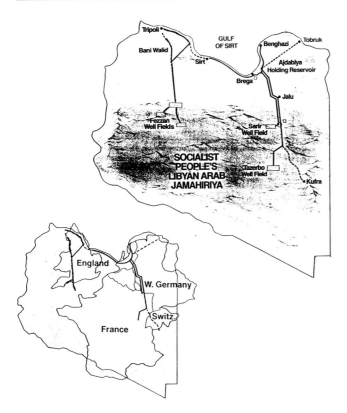

Figure 5.3 Libya's Great Man-made River Project and its scale, as shown in the overlay over France and Germany.

(some 600 km) distances up to the agricultural coastal areas. Additionally, it will provide water for industrial, domestic and municipal use. It is claimed that its use in the desert areas overlying these water resources would be uneconomical, and that conveying the water from the desert to the coastal region is more economical than any other alternative use (Salem, 1992).

The well-fields for the project are being constructed 400 to 700 km inland to tap the better quality water available there. The well-fields are spread over large areas where the aquifers come close to the surface.

The project development was planned to occur in separate phases. The first phase, the largest, was inaugurated 28 August 1991, and consists of a system that is designed to extract and carry up to two million m^3/day (23 m^3/s) to the coastal region where the majority of the population lives. The current plan is to expand this system to carry up to 3.68 million m^3/day sometime in the future, using a total of about 1,900 km of pre-stressed concrete cylinder pipe, ranging between 1.6 m in diameter for well-field networks and 4.0 m in diameter for the main conveyance pipeline, laid and buried in a 6 to 7 m deep trench. The second phase opened up a well field in western Libya. Water from this phase was

flowing to Tripoli on September 1, 1996. The remaining phases are to further expand and extend the Great Man-Made River Project to more areas along the northern coast of Libya.

Currently, Libya's water demands exceed its supply, and hence the motivation for this Great Man-Made River Project. Ground water is responsible for more than 98% of the total water consumption in Libya. Agriculture accounts for over 85% of the ground water used. The country is not agriculturally self-sufficient, and it may never be, but to reduce this dependency they plan increased irrigation development. In addition, more water is needed for meeting municipal and industrial demands along the northern coast, and to recharge coastal groundwater aquifers that have been over-exploited and are experiencing saline water intrusion.

There are many questions concerning the economic feasibility of the proposed future phases of this project. In addition there are potential water quality issues to be addressed. Clearly this source of ground water cannot last forever, at least not at reasonable costs. Other options to increase the long-term sustainability of Libya's water resource systems will have to consider increased water reuse and use efficiencies, desalination, and measures to improve agricultural security, if not self sufficiency. Right now, however, the Great Man-Made River Project appears to be one of the most cost-effective ways to supply Libya's increasing water demands. It can do this for quite some time, as the aquifers appear to have virtually unlimited (> 50 years) supplies. If the economic and environmental benefits exceed the costs, it well may be the most sustainable decision that can be made given current technology and costs. This may be true even though the water being pumped out of the desert is clearly not being replenished at the same rate.

Restoration of the Rhine Rivers (Europe)

The Rhine River catchment area (Figure 5.4) is about 185,000 km^2, and includes portions of nine European countries. About 55 million people live in the Rhine River basin and about 20 million of those people drink the river water. The basin is characterized by intensive industrial and agricultural activities. Some 20 percent of the world's chemical industry is located in the Rhine River basin. The river is also one of the most important in the world for shipping (van Dijk *et al.*, 1994; van Dijk, Marteijn & Schulte-Wulwer-Leidig, 1995).

In the mid-1970s, the Rhine was called by some the most romantic sewer in Europe. In November 1986, a chemical

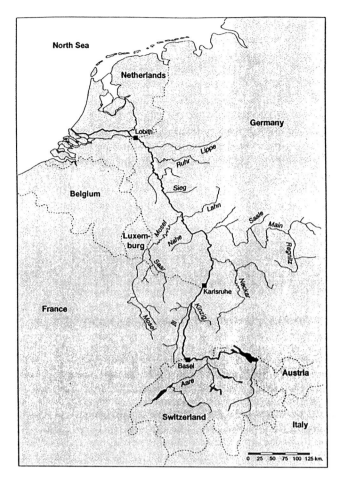

Figure 5.4 The Rhine River catchment in western Europe.

increasing realization among professionals and the public alike that natural rivers and their flood plains provide considerable economic as well as social benefits from recreation, flood control and water quality enhancement in addition to increased diversity of wildlife, including fish.

The mouth of the Rhine is in The Netherlands. The Dutch have set broad targets and have designed action programs to meet these targets. These targets translate into the following objectives:

- The Rhine must become a suitable habitat to allow the return of higher species (e.g., salmon) once found here.
- The use of the Rhine water for drinking water supplies must continue to be possible in the future.
- The river sediments must not be polluted by toxic substances.
- The North Sea environment, into which the Rhine flows, must be protected.

The objectives must be achieved by means of the measures included in the joint action program. They involve:

- accelerated reduction of permanent pollution caused by direct and diffuse emissions (e.g., a reduction target of 50 percent),
- protection against hazardous accidental emissions by companies situated along the river,
- improvement of the hydrological and morphological conditions.

It would appear that the current actions with respect to the restoration of the Rhine River and its basin are following many of the guidelines set forth in the beginning of the chapter. These decisions are not purely for environmental reasons, but for economic as well. Both criterion are relevant to sustainability.

Restoration of the Danube River Basin (Europe)

The Danube River basin (Figure 5.6) is in the heartland of Central Europe. The basin includes to a large extent the territories of twelve countries and collects additionally the runoff from small catchments located in four other countries. About 90 million people live in the basin. This river encompasses perhaps more political, economic and social variations than any other river basin in the world – certainly in Europe.

The river discharges into the Black Sea. The Danube delta and the banks of the Black Sea have been designated a Biosphere Reserve by UNESCO. Over half of the Delta has been declared a 'wet zone of international significance.' Throughout its length the Danube river provides a vital resource for drainage, communications, transport, power

spill destroyed much of the upper Rhine's aquatic ecosystem. One year later the Rhine Action Plan was initiated. Its mission is to restore the entire Rhine River ecosystem. This restoration program is commonly referred to as Salmon 2000 - reflecting the goal to restore the habitat for salmon, that once flourished in the entire Rhine, by the year 2000. Additional objectives include the provision of safe drinking water and the improvement of the quality of the sediments.

The shape (geomorphology) of the Rhine River has changed considerably during the previous century of canalization and other water resource development projects. Figure 5.5 illustrates this. These were mostly aimed at increasing the efficiency of river transport and hydropower production, and at increasing the protection from flood flows. This led to substantial ecological changes in the Rhine, e.g., migratory fish such as salmon lost their spawning grounds or could no longer reach them.

Two recent major floods on the Rhine have convinced everyone (at least for a while) that dirt, rocks and concrete alone will not provide sufficient flood protection. There is an

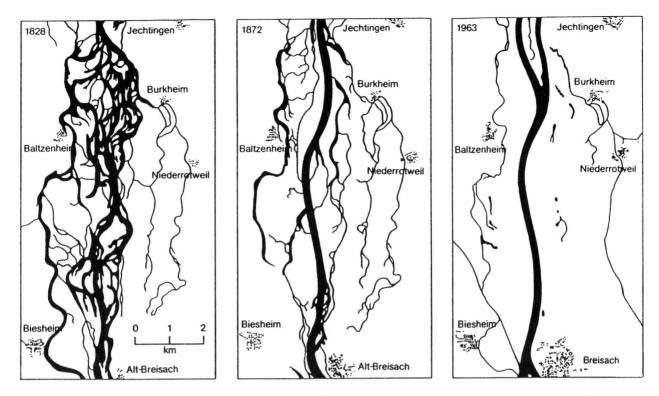

Figure 5.5 The canalization of a portion of the Rhine River over time.

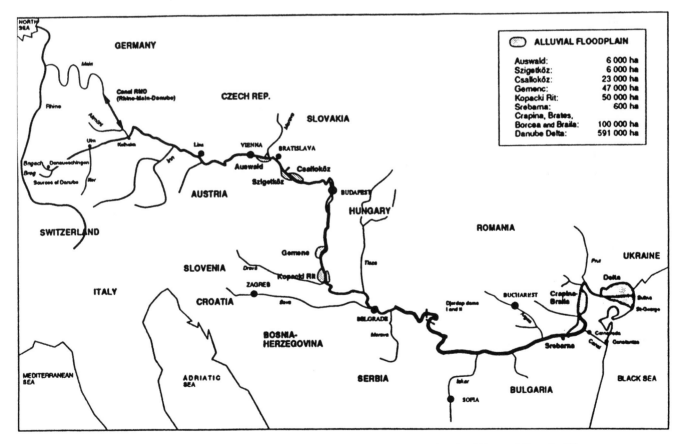

Figure 5.6 The Danube River Basin in central Europe. Also shown is the canal connecting the Rhine and Danube Rivers.

generation, fishing, recreation and tourism. It is considered to be an ecosystem with irreplaceable environmental values.

More than 40 dams and large barrages plus over 500 smaller reservoirs have been constructed on the main Danube River and its tributaries. Most of the length of the main stem of the Danube River and the major tributaries are confined by flood control dikes. Over the last 50 years natural alluvial flood plain areas have declined from about 26,000 km^2 to about 6,000 km^2. There are also significant reaches with river training works and river diversion structures. These structures trap nutrients and sediment in the reservoirs, causing changes in downstream flow and sediment transport regimes that reduce the ecosystems' habitats both longitudinally and transversely, and decrease the efficiency of natural purification processes. Thus while these engineered facilities provide important opportunities for the control and use of the river's resources, they are also illustrative examples of the difficulties of balancing these important economic activities with environmentally sound and sustainable management.

The environmental quality of the Danube River is also under intense pressure from a diverse range of human activities, including point source and non-point source agricultural, industrial and municipal wastes. Because of the poor water quality (sometimes affecting human health) the riparian countries of the Danube river basin have been participating in environmental management activities on regional, national and local levels for several decades (European Bank for Reconstruction and Development, 1993). A formal Convention on Cooperation for the Protection and Sustainable Use of the Danube River was signed by all Danube countries in June 1994. The countries have agreed to take '... all appropriate legal, administrative and technical measures to at least maintain and improve the current environmental and water quality conditions of the Danube river and of the waters in its catchment area and to prevent and reduce as far as possible adverse impacts and changes occurring or likely to be caused.'

North and Baltic Seas pollution (Europe)

The North Sea (Figure 5.7) is the most densely navigated sea in the world. But besides shipping usage there is the offshore

Figure 5.7 The Baltic and North Seas in northern Europe.

oil industry, telephone cables covering the sea bed and there is military and recreational use. It is a rich and productive sea with resources which include not only fish but crucial minerals (in addition to oil) such as gas, sand and gravel. These resources and activities are important to the surrounding countries for they play a major part in their economies.

As a sea so intensively used and surrounded by advanced industrialized countries which produce large waste flows, pollution problems are serious. The main pollution sources include rivers and other outfalls, dumping by ships (of dredged materials, sewage sludge and chemical wastes) and operational discharges from offshore installations and from ships. Atmospheric pollution entering the region is another major cause of pollution.

Those parts of the sea at greatest risk from pollution are where the sediments come to rest, where the water replacement is slowest and where nutrient concentrations and biological productivity are highest. Included in these are The Netherlands coastal waters (particularly the Wadden Sea), the German Bight and parts of the sea off the Danish west coast and Norwegian southern coast. The central and northern areas of the North Sea, are currently relatively free from pollution.

A number of warning signals have been received from the North Sea in its reactions to the stresses caused by the pollution. These include:

- Algal populations have changed in number and species. There have been large algal blooms, caused by excessive nutrient discharge from land and atmospheric sources.
- Species changes show a tendency toward more short lived species of the opportunistic type and a reduction, sometimes drastic to the point of disappearance, of some mammals and fish species and the sea grass community, polychaetes, coelenterates and meiofauna. A decrease of ray, mackerel, sand eel and echinoderms due to eutrophication have resulted in reduced plaice, cod, haddock and dab, mollusk and scoter. The negative impact of eutrophication could be due to enforced competition. Sole is the only fish species that benefits from eutrophication.
- The impact of fishing activities is also considerable. It has been shown that fish stocks are mainly affected by fisheries, whereas mammals and birds are mainly affected by pollution and zoobenthos (mainly by eutrophication).
- Particular concern has been expressed about the Wadden Sea, which plays a nursery function for many North Sea species. PCB contamination, for example,

almost caused the disappearance of seals in the 1970s. Also, the 1988 massive seal mortality in the North and Wadden Seas, although caused by a viral disease, is still thought by many to have a link with marine pollution.

Although the North Sea is sick enough to need radical and lengthy treatment it is probably not a terminal case. Actions are being taken by bordering countries to reduce the discharge of wastes into the sea. A major factor leading to agreements to reduce discharges of wastewaters has been the verification of predictive pollutant circulation models of the sea that identify the impacts of discharges from various sites along the sea boundary.

The Baltic Sea (Figure 5.7) is similar to the North Sea in that it is surrounded by industrialized countries and is close to the major sources of air and water pollution in Northern and Eastern Europe (HELCOM, 1993). It is subject to eutrophication from human activities in the Baltic drainage basin. Although over-fishing is clearly a major problem, decline in cod stocks and changes in the composition of near-shore fisheries can also be linked to the eutrophication process and its effects on regulator ecological functions and the services they provide.

Increased nutrient loads from agricultural runoff, municipal sources and atmospheric fallout have increased phytoplankton primary production by some 30 percent. This has broadened the base of the trophic pyramid. The resulting sedimentation of organic matter has depleted oxygen and benthic fauna, whereas the animal biomass on bottoms less than 50 m deep has increased four times since the 1920s. One implication is the significant decline in the food for cod in the Northern Baltic. Another is a reduction in the area that cod find suitable for reproduction.

In addition to the eutrophication is the widespread release of toxins and pollutants that accumulate in the sediments and in the food web. Sea mammals, sea birds and Baltic fish species have been particularly affected, and some animals, such as the gray seal and the sea eagle, are threatened with extinction.

The Orangi Pilot Project of Karachi (Pakistan)

In the early 1980s Akhter Hameed Khan, a well respected community organizer, began working in the slums of Karachi (Jordaan et al., 1993). Asking what problem he could help solve, he was told that 'the streets were filled with excreta and wastewater, making movement difficult and creating enormous health hazards.' 'What did the people want, and how did they intend to get it?' he asked. The reply? A traditional sewerage system . . . it would be difficult to get them to finance anything else.' And how they would get it, too, was clear – they would have Dr Khan persuade the Karachi Development Authority (KDA) to provide it free, as was done (or so the poor perceived) for the richer areas of the city.

Dr Khan spent months going with representatives of the community to petition the KDA to provide the service. When it was clear that this would never happen, Dr Khan was ready to work with the community to find alternatives. He would later describe this first step as the most important thing he did in Orangi – liberating, as he put it, the people from the immobilizing myths of government handouts.

With a small amount of core external funding, the Orangi Pilot Project (OPP) was initiated. It was clear what services the people wanted; the task was to reduce the costs to affordable levels and to develop organizations that could provide and operate the systems. On the technical side, the achievements of the OPP architects and engineers were remarkable and innovative. Thanks partly to the elimination of corruption and the provision of labor by community members, the costs (for an in-house sanitary latrine, house sewer on the plot and underground sewers in the lanes and streets) were less than $50 per household.

The related organizational achievements are equally impressive. OPP staff members have played a catalytic role: they explained the benefits of sanitation and the technical possibilities to residents, conducted research and provided technical assistance. The OPP staff never handled the community's money. The total costs of the OPP's operations amounted, even in the project's early years, to less than 15 percent of the amount invested by the community. The households' responsibilities included financing their share of the costs, participating in construction, and electing a 'lane manager' who typically represents about fifteen households. Committees, in turn, elect members of neighborhood committees (typically representing about 600 houses), which manage the secondary sewers.

The early successes achieved by the project created a 'snowball' effect, in part because of the increased value of properties with sewerage systems. As the power of the OPP-related organizations increased, they were able to persuade the municipality to provide funds for the construction of trunk sewers.

The Orangi Pilot Project has led to the provision of a sewerage system in Karachi, and to recent initiatives by several municipalities in Pakistan to follow the OPP method and (according to OPP leader Arif Hasan) have government behave like an NGO. Even in Karachi the mayor now formally accepts the principle of 'internal' development by the residents and 'external' development (including trunk sewers and treatment) by the municipality.

This study illustrates a bottom-up approach involving interested stakeholders who reap the benefits and for the most part, pay the costs. These conditions are usually a prerequisite for sustained system development, management and maintenance.

An ecosystem view for managing the Great Lakes basin (Canada and USA)

The Great Lakes (Figure 5.8) form a part of the border between Canada and the USA. Since they are international bodies of water, their management and 'health' are overseen by a number of commissions – one of which developed an Ecosystem charter for the Great Lakes Basin. This charter is to guide future actions to enhance and sustain the environmental health and economic viability of the world's greatest freshwater system. They hope it will serve as a model on how to manage other watersheds and river basins world-wide.

The Great Lakes Ecosystem Charter is a statement of shared principles and commitments for an array of stakeholders. Their definition of ecosystem includes the interacting components of air, land, water and living organisms (including humans) in the basin. Those who agree to operate and function in accordance with the charter principles will have a host of multiple objectives to achieve. Some of these, related to sustainability, include:

- An ecosystem approach to protecting, restoring and sustaining ecological processes and resources of the Basin Ecosystem shall be adopted. It will be predicated on the understanding that human activities, natural resources and ecological processes are parts of a unified whole and completely interdependent.

- Contamination of the Great Lakes Basin Ecosystem shall cease to the maximum extent practicable. This will be accomplished by eliminating or reducing the discharge of conventional pollutants, prohibiting the discharge of toxic substances in toxic amounts and virtually eliminating the discharge of all persistent toxic substances into the water, air and soil of the Great Lakes Basin Ecosystem.

- The Great Lakes Basin Ecosystem shall support self-sustaining communities of renewable resources that provide an optimum contribution of harvestable resources, recreational opportunities and associated benefits to meet societal needs for a healthy ecosystem,

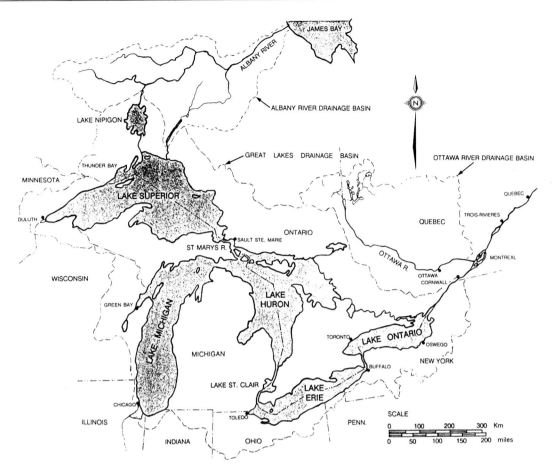

Figure 5.8 The Great Lakes Basin in North America.

wholesome food, raw materials, recreation and economic livelihood.

- Signatories shall recognize the interdependence between the health and integrity of the ecosystem and the economic well-being of human communities. They shall take the necessary measures to assure that protecting ecosystem integrity is an integral part of economic activity within the basin.
- Industry shall support and maintain high environmental, conservation and safety standards and principles.
- Public agencies, private enterprises and individual citizens shall recognize energy conservation as a priority necessary for the economic and ecological well-being of the Great Lakes Basin Ecosystem.

It will be interesting to see if such a Charter changes human behavior in the basin. The objectives specified in the principles are not specific, although a few could be. Thus it will be hard to judge whether they are being followed or not. Furthermore, it will be impossible to satisfy all the objectives, even if they were quantified, since some are conflicting.

Clearly tradeoffs must be made even among the subset of principles listed above. However, their very existence may cause people and organizations to think more about the impacts of their activities on the ecosystem as a whole, and the tradeoffs that are required between environmental, ecological, economical and social objectives that they have identified for their region, for their organizations and for themselves.

Flood management on the Senegal River (Senegal, Mauritania, Mali)

As on many rivers in the tropical developing world, dam constructions on the Senegal (or rather, conventional dam management strategies) can change not only the riverine environment but also the social interactions and economic productivity of farmers, fishers and herders whose livelihoods depend on the annual flooding of valley bottomlands. Although much of the Senegal river flows through a low rainfall area, the annual flood made possible a rich and biologically diverse ecosystem that supported dense human and animal populations. Living in a sustainable relationship with their environment, small-holders farmed sandy uplands during the brief rainy season, and then cultivated the clay plains as floodwaters receded to the main channel of the river. Livestock also profited from the succession of rain-fed pastures on the uplands and flood-recession pastures on the plains. Fish were abundant; as many as 30,000 tons were caught yearly. Since the early 1970s, small irrigated rice schemes added a fifth element to the production array: rain-fed farming, recession farming, herding, fishing and irrigation.

Completion of the Diama salt intrusion barrage near the mouth of the river between Senegal and Mauritania and Manantali High Dam more than 1,000 km upstream in Mali, and the anticipated termination of the annual flood due to the dam (because of common belief that releasing large quantities of water to create an artificial flood is incompatible with maximum hydropower production) will have adverse effects on the environment. Rather than insulating the people from the ravages of drought, the dam release

policy can accelerate desertification and intensify food insecurity. Furthermore, anticipation of donor investments in huge irrigation schemes has, in this particular case, led to the expulsion of non-Arabic-speaking black Mauritanians from their floodplain lands.

This is a common impact of dam construction: increased hardships of generally politically powerless people in order that urban and industrial sectors may enjoy electricity at reduced costs. Such impacts are often unnecessary without much loss in energy production.

Studies in the Senegal Valley (Figure 5.9) by anthropologists, hydrologists, agronomists and others suggest that it may be entirely feasible to manage the dams with an annual controlled release, and 'artificial flood,' assuring satisfaction of both urban, industrial and rural demands for the river's water and supporting groundwater recharge, reforestation and biodiversity.

Because of these studies, the Sengelese government ended its opposition to an artificial flood, and its development plans for the region are now predicated on its permanence. As of

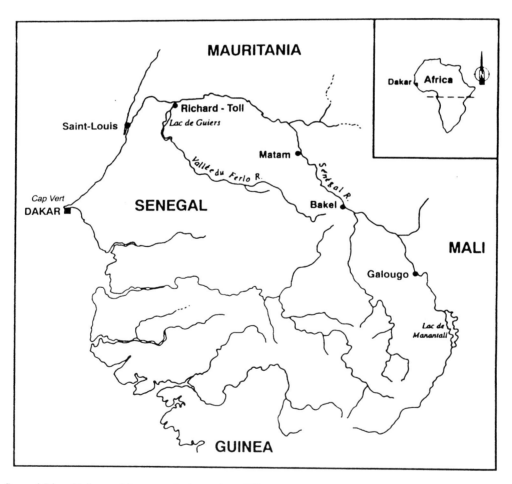

Figure 5.9 The Senegal River Valley and its reservoirs in western Africa.

late 1993, however, the other members of the three-country consortium managing the dams – Mali and Mauritania – remained committed to the original scenario (McDonald & Kay, 1988; Horowitz, 1994)

Sedimentation of Tarbella Dam (Pakistan)

Tarbella Dam, on the Indus River in Pakistan is the largest earth-fill dam in the world. It was built by large earth-moving equipment. This happened in a region having thousands of unemployed or under-employed workers. The high technology approach to dam building was selected because the low bidder did not feel confident in its ability to organize and manage large numbers of workers. While mobilizing and managing very large numbers of workers can and has been done elsewhere, it is just not as convenient to use labor-intensive technology when mechanical capital-intensive technology is available (and there are no social security costs involved).

Regardless of how it was constructed, the dam is projected to fill with sediment within 40 years of its life. After that it will be a run-of-river project. Will the reduced benefits of a run-of-river project compared to a variable head reservoir mean that future generations of users of that water resource system will wish the dam had not been built? When will the economic life of the dam end, and will the total benefits derived during its economic life outweigh the costs, if any, of having a run-of-river dam to operate and maintain, or destroy? And what are its environmental and ecological impacts over time? All these questions are relevant when considering whether or not the decision to build this dam contributed to the region's sustainability.

Aesthetics and the Olympus Dam (USA)

Olympus Dam in Estes Park, Colorado, has been a highly successful structure from the viewpoint of delivering much needed water to the irrigated lands and the cities which are west of the continental divide in Colorado. From the viewpoint of aesthetics, however, it is a mass of concrete and steel and a long bare artificial rock dike in the foreground of a beautiful mountain range and valley to be seen by any visitor approaching the area from either of the two eastern entrances. A site was (and still is) available just a short distance downstream for a much smaller dam that would have accomplished the same purpose, would have been less expensive and would have been hidden from view. Political pressure, however, forced the selection of the existing aesthetically unattractive site instead of the less expensive and hidden site for the dam.

If the project were being planned today, it might very likely be opposed by those who value, and can obtain popular support for, beauty (Albertson, 1995). In this case political pressure came from those who were not interested in aesthetics, or at least were not forced to consider this aspect of sustainability. Today a more bottom-up stakeholder involvement might have resulted in a different decision with respect to dam location and design.

Public pressures and the Central Utah Project (USA)

The Central Utah Project (CUP) is a large water resources development project including irrigation and hydropower. The US Bureau of Reclamation initiated this project through a local Utah organization – an organization that was later found not to represent the people being served. After the project was begun and well underway, construction was stopped for lack of government funding. The US Congress refused to continue to appropriate funds because of a very vocal opposition to the CUP by citizen groups that formed. Under the leadership of a US Senator and the local groups, the project was eventually reorganized with new plans and designs created by private engineering companies.

This is an example of a project that was modified because of public – stakeholder – pressure. Increasingly it is being recognized that the planning and designing processes cannot proceed in an environment that excludes public participation in these processes.

Water supplies in Tchelo Djegou (Niger)

Tchelo Djegou is a village of about 200 semi-nomadic herders located in a remote region of rural Niger in West Africa. Niger, located in the southern region of the Sahara Desert, has an extremely hot and dry climate. Most regions of the country receive less than 800 mm/year of rainfall, all of which falls during a brief three-month rainy season. The village, and the country have suffered severe droughts during the past 30 years, killing many of the nomadic population. Walking up to 10 km to collect water for daily needs is not uncommon.

As in many of the surrounding communities, a 110 m well and mechanical hand pump were installed in Tchelo Djegou as part of a region-wide international water resources development project in the early 1970's. The new water supply attracted those families that were forced out of the river valley in search of new fields, and the village population increased. As the pump aged, however, maintenance costs increased and eventually exceeded the financial resources of the village. The pump was abandoned and the villagers were forced to make an exhausting walk to a distant river valley with pails carried on top of their heads.

In the fall of 1991 the village elected to implement a more appropriate and affordable technology. After experimenting with several alternatives, a pump was fabricated from available materials for a total cost of less than $1,200. A wheel, 3 m in diameter, acts as a spool onto which 100 meters of nylon cord are wrapped. A modified version of a bailer is attached to the cord, and is centered over the well opening with a frame and pulley. The system is operated by rotating the wheel, usually requiring three people, and in three minutes can deliver a pail of water from over 94 m below the surface. The village's total water supply can be met using the well and pump for about 5 hours per day.

Will population increase result in increased water demands beyond the well's capacity? Time will tell, but clearly the cost of maintaining this well appears to be far less than that of the older mechanical well (Brown, 1995).

This case study is just one of many small projects that do not result in large dams or water treatment and distribution systems, and do not involve large amounts of investments and technical support. They do, nevertheless, affect the economic and social sustainability of the local region.

Hydropower dam construction in Malaysia

A six-billion dollar dam, under construction in Malaysia's eastern Sarawak state, will be Southeast Asia's biggest (after China's Three Gorges Dam) when it is completed. It will inundate an area the size of Singapore. Not only does it require clearing nearly 70,000 hectares of tropical forests in Sarawak, but an environmental impact assessment on the project indicates that the wildlife, lifestyles of some 5,000 people (mostly farmers) and the biodiversity of a surrounding 1.5 million hectares will be affected. The dam's catchment area supports more than 800 plant species, of which 67 have protected status, and 229 mammals and bird species, of which 43 are protected.

As opposed to the environmental impacts the energy will be produced by a 2,400 megawatt hydroelectric power plant. The Malaysian government considers the dam essential for the nation's energy demands. But, of the total, only 1,000 megawatts will be used in Malaysia, the remainder going to Indonesia and Thailand through submarine cables. Clearly there are costs and benefits associated with any decision that may be made. There are also long-term as well as short term

economic, environmental, ecological and social impacts that are relevant to any consideration of sustainability.

It seems the only ones who are in a position to judge the merits of this project, in all its aspects, are those directly affected by it. In a project such as this where it may not be obvious what decisions are the right ones, all concerned stakeholders must be involved and be heard by those who are responsible for making these decisions.

High Aswan Dam revisited (Egypt)

Today only a small fraction of the total Nile River flow reaches the Mediterranean Sea. Egypt's agricultural production remains limited by water, not land. The country's population has grown to the point where now the arable land per capita is among the lowest of any country in Africa. If it were not for the construction of the High Dam in 1968, this amount might be even lower. Because of the High Aswan Dam, the total arable land in the Nile Valley, in Egypt, increased from 6 million feddans (1 feddan is equal to 1.038 acres or 0.415 ha) in 1970 to about 7.2 million feddans in 1990, a twenty percent increase. This, together with the increased water supply reliability, hydropower production, navigation and flood control has made this dam an economic success.

Downstream from the dam, there have been obvious physical and environmental impacts, for the dam has changed the hydraulic regime of the river. It is now a glorified canal whose flow is totally controlled by a series of dams and barrages. Canalization of the river has reduced its length. The number of islands in the river has declined from 150 to 36, at present. River-bed erosion has resulted in a rather modest 0.25 m drop. The reduction of natural deposition of silt on the downstream flood plains during the annual flood is now compensated for by the addition of some 13,000 tons of lime-nitrate fertilizer. Coastal erosion has occurred, as predicted. Over-irrigation and other inefficient uses of water and crop production practices have contributed to an increase in the groundwater levels and in turn, increased salinity and water-logging. Fish catch has been reduced to about half of what it was before the dam was built. Provision of clean water and sanitation, health education and rural clinics has reduced the overall prevalence of schistosomiasis from more than 40 percent during the pre-dam period to only 10.7 percent in 1991.

Adaptive management on the Columbia River (USA)

The Columbia River (Figure 5.10) is the fourth largest river in North America. As with other large water and related

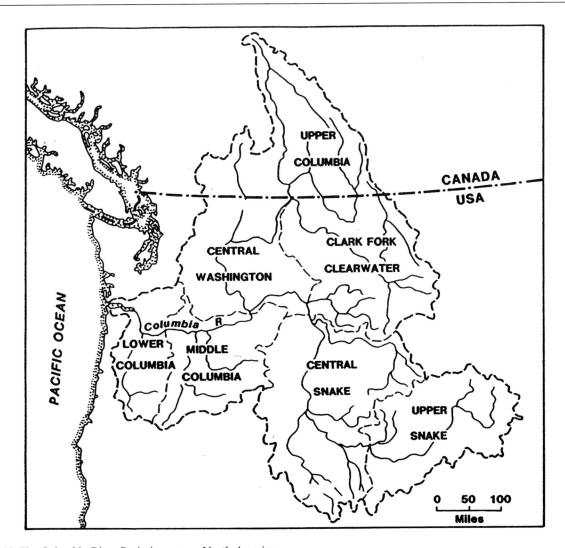

Figure 5.10 The Columbia River Basin in western North America.

ecosystems, its resources and problems involve a number of regional governments and conflicting uses. When hydropower dams were constructed on the river, intense development of the region followed, with significant environmental costs.

Predevelopment salmon populations approximated 11 million adults annually, of which 77 percent used the upper river for spawning. In the early 1980s spawning runs averaged only 3 million per year, with only 42 percent using the waters above the lower Bonneville Dam. Today, two-thirds of Columbia River salmonids come from hatcheries. This decline in salmon not only represents depleted resources, to many it has degraded the very spirit of the Pacific Northwest. Additional stresses on the Columbia River ecosystem include fishing and agricultural water uses.

In August, 1995, seven public and private partners announced a $1.4 million project to acquire 229 acres and enhance and restore approximately 3,500 acres of wetlands, riparian forest, and upland fish and wildlife habitats along the lower Columbia River. The partners have taken an ecosystem approach to management. They seek to restore wetlands and develop water delivery systems to them that will allow the agencies to mimic the natural flood regimes that occurred prior to construction of dams upstream on the Columbia River and its tributaries. Project work includes restoration or enhancement of both permanent and seasonal wetlands, installation of electric pumps to provide dependable year-round water supplies, reconstruction or restoration of water delivery systems, construction of fences to protect riparian forests from livestock grazing, planting of trees, shrubs, and other vegetation, and improving grasslands adjacent to wetlands.

All of these improvements are designed to provide habitat for migrating and wintering waterfowl, such as ducks,

Canada geese, and tundra swans. The lower Columbia River wetlands support the second largest migrating and wintering populations of waterfowl on the Pacific Northwest coast, numbering over 200,000 birds. These habitats also support more than 250 species of other birds, including resident and migratory bald eagles, hawks, shorebirds, waterbirds, and songbirds. Pacific salmon also will benefit through the improved Columbia River water quality provided by these wetlands.

By improving waterfowl habitat on Federal, state, and county lands, the agencies intend to reduce crop depredations caused by wintering geese on nearby farm lands. The improved habitats, by attracting and holding more wildlife, will improve recreational and educational opportunities for visitors. The public lands are popular sites for wildlife observation, environmental education, photography, waterfowl and upland game bird hunting, and fishing.

The lower Columbia River ecosystem contains a unique complex of freshwater tidal systems that include marshes, forests, shallow lakes, ponds, sloughs, and grasslands. In addition, heavy winter rains provide seasonal flooding of upland marshes, swales, and farm lands. Scientists estimate that more than three-fourths of the wetlands that historically existed along the lower Columbia between Bonneville Dam and Astoria, Oregon, have been converted to agriculture, urban and industrial uses, and other activities, such as dredge spoil disposal and flood control.

In moving toward a sustainable future, human interaction with such complex ecosystems is being guided by adaptive management (see Chapter 7) and alternative dispute resolution methods. Adaptive management is an approach to natural resource policy that embodies a simple imperative: policies are experiments, learn from them. Since we do not understand enough about how to manage such ecosystems adequately, a trial and error learning approach is used. Results are monitored and evaluated, and corrective actions are taken if warranted. This approach integrates management and acquisition of new knowledge in the longer time frames and larger spatial scales needed as managers grapple with complex problems, such as the restoration of salmon habitats in aquatic ecosystems.

Restoration of the Kissimmee River (USA)

Beginning in 1951, a channelization project for flood control converted 166 km of the natural meandering channel of the Kissimmee River in central Florida (Figure 5.11) into a 90 km canal. Excavation of the canal and associated deposition of spoil and de-watering of the floodplain destroyed 56 km,

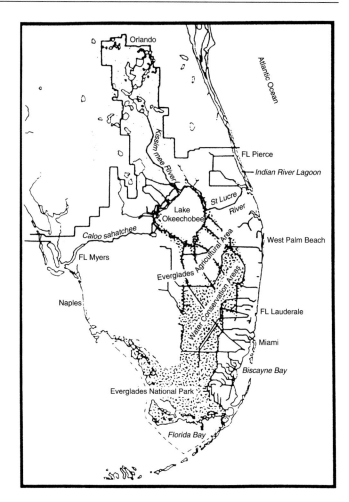

Figure 5.11 The Kissimmee River, Lake Okeechobee and the Everglades area in central and southern Florida.

or 34 percent, of the original river channel and 2,800 hectares, or 14 percent of the original floodplain wetlands. Habitat degradation and declining water quality generated widespread public concern over loss of natural resource values soon after the project was completed in 1971. Various acts were subsequently passed in the state legislature to restore the river and its watershed.

A complex pilot restoration project funded by the state of Florida and the applicable water management district demonstrated the feasibility of restoring the integrity of the combined river and floodplain ecosystems. Back-filling of the canal and installation of weirs to redirect flow from the canal into selected reaches of the old river channel are re-establishing the mosaic of habitats and of the hydrologic regime to patterns of variation that existed before canalization. As a result, habitat complexity and diversity have increased, wetland vegetation has recovered in channel and riparian areas, populations of fish and waterfowl have increased and water quality has improved.

The success of this pilot project has, in part, provided the basis for implementing the restoration of the entire Kissimmee River, including back-filling about 56 km of the canal. This back-filling will restore flow to the channel and reconnect the channel to its original floodplain. This project is a joint Federal, State, and local water management district effort, and provides an excellent example of interagency cooperation and coordination (South Florida Water Management District, 1988).

The Hula Valley Project (Israel)

Hula Lake and marshes are part of the watershed of the Sea Of Galilee (Lake Kinneret) in the Jordan Valley. During the early settlement of Israel in the late 19th and early 20th centuries, the lake was considered a curse because of the mosquitoes in its shallower waters and the malaria they spread. However the lake and the adjacent marshes were also a rich ecosystem, with hundreds or thousands of species of flora and fauna, large and small, including water buffalo and papyrus plants, some found exclusively in this area. The valley has peat soils that were very fertile.

A project was initiated to 'conquer the swamp', dry up the land, convert it to productive agriculture, and eradicate the malaria bearing mosquitoes (Dimentman & Bromley, 1992). Planners, engineers and the public enthusiastically supported the project. The minority who were concerned about draining peat soils and destroying the ecosystem, were considered irrelevant. The grand objective was more land and less malaria. Canals were dug, large areas were drained, and one small 'nature' area was preserved in one corner of the valley. The remaining drained area was converted to agriculture.

With time problems became apparent. Land subsidence was very pronounced, causing difficulties in cultivation. The dried peat oxidized, and fires erupted spontaneously. Nutrients began leaching from the dried up area and traveled downstream into the Sea of Galilee, one of the country's major sources of potable water, causing concern for eutrophication. The nature reserve in the Hula Valley was criticized by ecologists as being very different from the ecosystem it was supposed to preserve.

These problems led to a reevaluation of the project in the late 1980s. By this time agriculture had lost its prime position in Israel, and farmers were looking for alternative economic activities in the area, primarily domestic country-style tourism. The study of the situation and of alternative plans established that there were three main objectives: (1) preservation (in fact, restoration) of the original natural ecosystem in part

of the area, (2) agriculture, and (3) alternative income-producing activities, primarily domestic tourism. The quality of the water draining from the area to the Sea of Galilee had to be maintained at levels which protect the quality of this lake.

A multi-objective study was conducted, with extensive involvement of the stake-holders. These included local residents, farmers, those interested in developing tourism, the 'green' groups, and representatives of the water authority responsible for the Sea of Galilee. The study ended with adoption of a new plan that is a compromise solution, as it always is in multi-objective cases. Part (about 6%) of the area which was dried in the original project was again flooded and a new nature reserve created. New canals and reservoirs were constructed, the drainage of the area was changed to maintain moisture in the peat soils, and a new management scheme was implemented. The new project was inaugurated in 1994, and it remains to be seen whether it is a greater success than the original project.

Was the original plan sustainable? Is the current plan likely to be? A few comments are offered. The original objective of agricultural development, combined with malaria eradication, overrode all concerns with respect to the ecosystem. Ignorance with respect to the peat soils, combined with an unwillingness to listen to advice that did not support the popular idea, led to a project that was later regretted. But maybe this after all is not a case of non-sustainable development. For 40 years the dried area served for agriculture, and now that its value is reduced and the negative consequences of the project are evident, a new plan has taken the place of the old one. Evolution of a plan during its life cycle may be taken as a sign of robustness, of adjusting to changing objectives and circumstances and to the emergence of new understanding and new data. The new replaces the old. Preservation of the ecosystem, tourism, some – but not total – agriculture, new information about management of peat soils, concern for the quality of the Sea of Galilee, increased involvement by all stakeholders in a process dominated by engineers and planners in the past, new methods of multi-objective planning and negotiation, are what is new today. Tomorrow today's understandings, objectives, and methods may be replaced with even newer ones and the plan may be modified accordingly.

Thus an evolving and adaptive plan may be a sign of sustainable development. Inflexibility in the face of new information and new objectives and new social and political environments is what may lead to reductions in project sustainability. In this case there is certainly regret regarding the decisions originally made. Had the planners taken into consideration the advice regarding the expected behavior of the peat soils, and the concerns regarding the preservation of the unique ecosystem (in which over 120 species of fauna were lost due to drying), they would probably have come up with a plan with less negative effects, one that could be adjusted more easily as objectives changed.

The Milwaukee story (USA)

Waterborne diseases are a product of inadequate sanitation in water resource systems. In 1993 some 370,000 citizens of Milwaukee, Wisconsin, were infected by an intestinal protozoan parasite, Cryptosporidium. Many lost work time and several deaths were associated with the outbreak. The contamination was traced to oocysts of the protozoan in the city's water supply, drawn from Lake Michigan. The protozoan has also infected water supplies in Georgia, Texas, New Mexican and elsewhere. The Milwaukee story became 'the Washington, DC' story for several days in December, 1993. Control of protozoan infections is a continuing concern within the watersheds of the reservoirs supplying the nation's large cities. An improved understanding of the dynamics of the freshwater systems through which this and similar diseases are transmitted is essential for maintaining water quality and human health.

Today, New York City, a much larger city than Milwaukee, gets its water, essentially untreated, from watersheds that support farming activities. Thus there is the possibility that Cryptosporidium could infect those drinking the water in New York City, and for this reason the US Environmental Protection Agency has ordered NYC to install water filtration plants, the cost of which would be in the billions of dollars. NYC has argued that this should not be necessary if they can show that they can control conditions in their watersheds that may cause the introduction of Cryptosporidium in their water supplies. This involves implementing farm management practices that can control the runoff from pasture lands and buying up watershed protection zones. Neither are popular with the local rural population. The US EPA has allowed NYC to prove that they can persuade the farmers to do what the city and their advisors believe necessary and to show that what is done is effective. If it works it may save substantial sums of money. Many believe, however, that in the long run, filtration, and its expense, will be necessary.

Damming the Mekong (SE Asia)

The Mekong River (Figure 5.12) flows 4,200 km through southeast Asia to the South China Sea through Tibet, Myanmar (Burma), Vietnam, Laos, Thailand and Cambodia. Its 'development' has been restricted over the past several decades due to regional conflicts, indeed conflicts that have altered the history of the world. Now that these conflicts are reduced, investment capital is becoming available to develop the Mekong's resources for improved fishing, irrigation, flood control, hydroelectric power, tourism, recreation and navigation. The potential benefits are substantial, but so are the environmental and ecological risks.

During some months of the year the lack of rainfall causes the Mekong to fall dramatically. Salt water may penetrate as much as 500 km inland. In other months the flow can be up to 30 times the low flows, causing the water in the river to back up into wetlands and flood some 12,000 km^2 of forests and paddy fields in the Vietnamese delta region alone. The ecology of a major lake in Cambodia depends on these backed up waters. While flooding imposes risks on some 50 million inhabitants of the Mekong flood plain, there are also distinct advantages. High waters deposit nutrient-rich silts on the low-lying farmlands, thus sparing the farmers from having to transport and spread fertilizers on their fields. Also, shallow lakes and submerged lands provide spawning habitats for about 90 percent of the fish in the Mekong basin. Fish yield totals over half a million tons annually.

What will happen to the social fabric and to the natural environment if the schemes to build big dams across the mainstream of the Mekong are implemented? Depending on their operation, they could disrupt the current fertility cycles and the habitats and habits of the fish in the river. Increased erosion downstream from major reservoirs is also a threat. Add to these possible adverse impacts the need to evacuate and resettle thousands of people displaced by the lake behind the dams. How will they be resettled? And how long will it take them to adjust to new farming conditions?

There have been suggestions that a proposed dam in Laos could cause deforestation in a wilderness area of some 3,000 km^2. Much of the wildlife, including elephants, big cats and other rare animals, would have to be protected if they are not to become endangered. A World Bank study suggests that the water would be unsuitable even for animals to drink some of the time. Malaria-carrying mosquitoes, liver fluke and other disease-bearers might find ideal breeding grounds in the mud flats of the shallow reservoir. Clearly these issues need to be considered now that development seems possible, and even likely.

In this context it is interesting to examine the impacts of a dam constructed on the Nam Pong River in northeast Thailand. The Nam Pong project was to provide hydroelec-

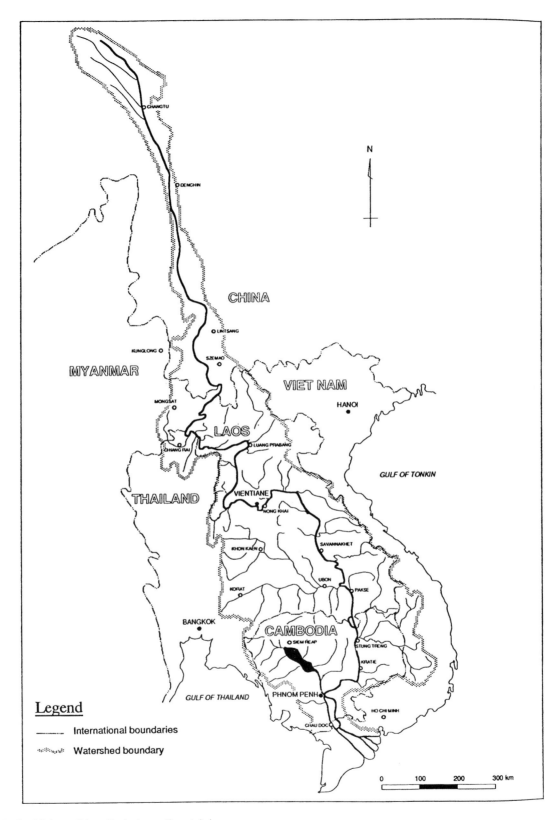

Figure 5.12 The Mekong River Basin in southeast Asia.

tric power and irrigation water, as are the avowed purposes of many reservoir projects throughout the world. Considerable attention was paid to the social aspects of this project, but not to the environmental impacts. The project had a number of unexpected consequences, both beneficial and adverse.

Because the reservoir was acting as a bio-reactor for most of the year, fish population became so large that a major fishery had developed in the reservoir. The economic benefits of fish production exceeded those derived from hydropower. However, lack of adequate planning for this event resulted in less than ideal living and economic conditions for the migrating fishermen who came to this region.

Despite the availability of irrigation water, most farmers were still practicing single-crop agriculture after the dam was built, and were still growing traditional crops in their traditional ways. No training was provided for them to adapt their skills to the new conditions and opportunities. In addition, while farming income did not decrease, the general welfare and health of the population seems to have decreased. Again, little attention was given to training about diet and hygiene under these new conditions.

The reservoir itself had some adverse impacts along with the beneficial ones. The adverse impacts included increased erosion of the stream banks, silting up of the channel and a large increase in aquatic vegetation that clogged hydraulic machinery and reduced transport capacity.

Volts from the Volta River (Ghana)

The Volta River dam creating Lake Volta in southern Ghana (Figure 5.13) was built in the early 1960s to provide power for the production of aluminum. The river drains almost all of Ghana as well as larger areas upstream. The Ghanaian government at the time the reservoir was built was committed to building the dam, regardless of environmental or ecological impacts, and the project has brought substantial economic benefits, including cheaper power. This in turn has enhanced the quality of life in the basin. There has also been a large increase in fish production within the reservoir and an accompanying large influx of migrant fishermen and their families. But these migrant fishermen and their families, living in scattered settlements along the lake shore, have degraded the local environment. Resettlement programs and agricultural programs associated with resettlement have not worked out very well. Reasons for this include farmers' resistance to change, over-dependence on the government, little enthusiasm for self-help, conflicts between old

and new populations, waterborne diseases and growth of aquatic vegetation (McDonald & Kay, 1988).

Conclusions

These experiences, when viewed through sustainability glasses provided by the suggested guidelines summarized at the beginning of this chapter, all point to at least two conclusions. One conclusion is that what we often think is best now for our current generation of water resource users may not be what we think may be best in the future. Current objectives and future objectives may not, in fact probably will not, be the same. Tradeoffs often must be made not only among today's objectives but also between what we think is a best compromise decision today and what we think our future generations may want. Another conclusion is that what we often think is best now turns out not to be as good as we had expected. Our ability to predict not only the various environmental, ecological, economical, financial, hydrological, and social impacts, but also what society will value or want in the future will always be inadequate. Hence system adjustments over time are needed.

All of these case studies point to the need for change over time. Our water resource systems must be designed and managed in a way that facilitates adaptation to changing environmental, economic and social conditions. What we have done in the past may not meet current objectives and concerns, and what we do now may indeed not be exactly what we or our descendants would wish we would have done. This does not necessarily mean the systems themselves have not fully contributed to society's objectives and cannot continue to do so on into the future. They simply have to adapt to those differing needs and objectives and conditions. It is our task today to ensure that our systems are robust, i.e., that they are designed and managed in a way that enables them to adapt to unforeseen changes at minimal environmental, economic and social costs.

To be able to support and adapt to changing societal objectives, the ecological, environmental and hydrological integrity of water resource systems must be maintained. The sustainability guidelines for designing, managing and operating the physical infrastructure; for protecting and enhancing the environment and ecosystems; for ensuring economic and financial feasibility; for considering impacts on institutions and society; for improving human health and welfare; and for implementing appropriate planning and technology are offered as ways by which planners and engineers can work to achieve this needed adaptability, and hence increased system sustainability.

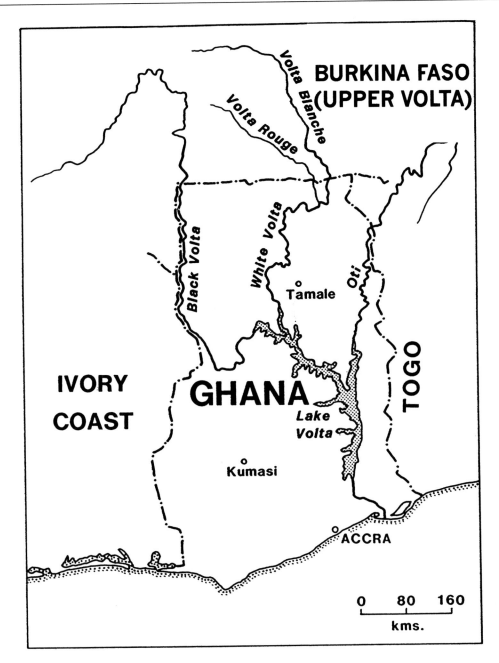

Figure 5.13 The Volta River Basin in western Africa.

SOME THINGS TO REMEMBER

- Review the sample guidelines for water resource systems engineers and planners: (1) for design, management and operation of physical infrastructure; (2) with respect to environment and ecosystems; (3) with respect to economics and finance; (4) with respect to institutions and society; (5) with respect to health and human welfare; and (6) with respect to planning and technology.

- In considering the case studies, recall that there were reasons for the decisions taken in each case, many of which we certainly do not and cannot know.

- When considered from a national perspective, some of the sub-regions might be degraded in an attempt to increase the probability that the remaining regions will become more sustainable. But this should be carefully considered before implementation.

- If the economic and environmental benefits exceed the costs, it may well be the most sustainable decision that can be made given current technology and costs. That could be true even though the water being transported from its source is clearly not being replenished at the same rate.

- A bottom-up approach involving interested stake-holders who reap the benefits, and for the most part pay the costs, is usually a prerequisite for sustained system development, management and maintenance.
- Increasingly it is being recognized that the planning and designing processes cannot proceed in an environ-ment that excludes public participation in these pro-cesses. All concerned stakeholers should be involved and be heard by those who are responsible for making the decisions.
- An evolving and adaptive plan may be a sign of sustain-able development. Inflexibility in the face of new infor-mation and new objectives and new social and political environments is what may lead to reductions in project sustainability.

6 Economic sustainability criteria

USE AND NON-USE ECONOMIC VALUES ——

Water and other environmental resources provide three main types of services to humans. First, these resources are essential inputs that support many economic activities. Second, these environmental resources serve as a sink to absorb and recycle some of the waste products of economic activities. Finally, they provide an irreplaceable life support function. All these services are valuable. Hence they can be considered economic services.

Animal and plant life is not possible without water. Our health depends on the quantity and quality of the water we drink. Even though we are completely dependent on water for life, however, this does not mean that water has an infinite value. What people pay for water depends in part on the cost of obtaining water where they are located, *and on how important it is for them to remain where they are*.

An inventory of all potential beneficial uses that can be provided by a particular water resource system is one of the first steps in any planning process. To the extent that the value of these beneficial uses can be expressed in monetary terms, they become economic uses and economic analyses can be used to identify the costs and benefits of alternative allocations to these uses. Monetary values indicate the relative economic desirability or the worth of a particular service, use, condition or impact.

Values derived from any water resource system can usually be divided into two types: use and non-use values. User benefits relate to the use to which water and water resource systems are put. User values can be subdivided into consumptive and non-consumptive values. Consumptive use values include those derived from allocations or diversions of water to consumptive water-using industries and agricul-

ture, or for cooling, or for waste transport. Once used, the water or its particular characteristic or property is not available for others to use. It is, therefore, not just the reduction in the amount of water that determines whether a use is consumptive or not. It is also the reduction of any quality characteristics of that water that otherwise could be beneficially used elsewhere. Changes in water quality characteristics that prevent its beneficial use at some sites may not preclude its beneficial use at other sites – such as the reuse of urban waste waters for irrigation of lawns, parks and golf courses.

Non-consumptive use values include the benefits received by those who leave the water and its characteristics essentially intact for others to use, such as most water-based recreation, beautiful scenery and wildlife. These non-consumptive uses are also called environmental amenity values. Some of these non-consumptive use values can be option values. Some individuals not wishing to use the water resource now may be willing to pay to preserve it for some future use. The amount they are willing to pay is described as an option value (Pearce, Markandya & Barbier, 1989).

Besides use values, there may also be individuals who desire and are willing to pay, or forgo current benefits, to preserve an asset for the benefit of future generations. These non-use values are sometimes called intrinsic or existence values. They are values placed on the mere existence of a resource and its physical, biological or cultural characteristics. These values are not associated with any specific use or set of uses. It is not always possible to separate non-consumptive and intrinsic values, and it is probably not all that important to do so. Figure 6.1 shows some of the relationships among these various economic values, and Table 6.1 lists some consumptive and non-consumptive uses of

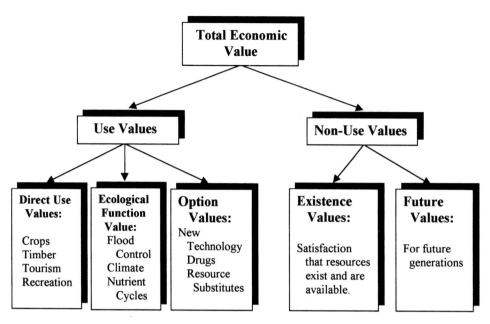

Figure 6.1 Various types of economic values sometimes assigned to water and other natural resources.

water resource systems that can be expressed in monetary terms. Table 6.2 identifies some amenity or intrinsic values of river systems (Delft Hydraulics Lab., 1994).

To estimate the monetary value of those water resource system functions, services or purposes that result in something that is produced and sold in the marketplace (such as crops from an irrigation area, or goods produced by industry, or electricity produced from hydropower, or cargo transported on waterways) is to perform benefit–cost analyses. A lower bound on the monetary value of non-marketed consumptive and non-consumptive uses (such as recreation, wastewater transport and assimilation and municipal water supplies) can be obtained by estimating what the minimum cost would be to obtain the service by another means, assuming those costs would be paid. Other approaches are also available, and are often grouped under the heading of environmental economics.

ENVIRONMENTAL ECONOMICS

Methods of environmental economics can help one identify environmental benefits and damages within comprehensive benefit–cost analyses of alternative systems. They include various concepts and techniques for estimating the monetary value of environmental impacts and for including (internalizing) these costs and benefits, in analyses for economic decision-making. Internalizing environmental externalities (i.e., the benefits or damages to others that are not received or paid by those that cause those benefits or damages) throughout the planning and development processes permits more effective and more detailed considerations of various alternatives. It also permits better and earlier opportunities for redesigning projects and policies to achieve sustainable development objectives.

The identification of sustainable development options requires, among other factors, an estimate of the economic value of any damage to the environment that any water resources system design or operating policy may cause. This is essential information for making environmentally sound investment decisions, but it is not always easy information to obtain. The remainder of this chapter outlines some of these techniques and identifies some modeling approaches that can help us estimate how to use our water resources over time, and how to deal with discount rates that tend to decrease the importance or significance of future events, especially distant future events, when compared to present or current events. (Alternative modeling approaches are plentiful. See for example those of Norgaard & Howarth, 1991; Munasinghe & Lutz, 1991; Munasinghe, 1993; Pezzey, 1992; Rotmans *et al.*, 1994, to list only a few.)

Environmental economics plays a key role in identifying options for efficient water resouce management. It provides a bridge between the traditional techniques and the emerging more environmentally sensitive approach of decision-making. Environmental economics helps us incorporate ecological concerns into the conventional economic decision-making framework.

Table 6.1. *Example of some consumptive and non-consumptive uses of water systems that may be valued in monetary terms, and the simulation variables that will determine how well the purpose is served. Functions of these simulation variables convert them to monetary values.*

Purpose served having monetary value	Simulation variables
Municipal and industrial supply	Water flows, storage volumes, quality
Agricultural water supply	Water flows, storage volumes, quality
Hydropower	Water flows, heads
Flood control	Water flows, storage volumes
Fishing	Water flows, storage volumes, velocities, quality
Recreation	Water flows, storage volumes, velocities, quality
Navigation	Water flows, elevations, depths, widths
Wastewater transport and assimilation	Water flows, storage volumes, quality
Sand, clay, gravel dredging	Water elevations, flows, sediment concentrations
Cooling water	Water flows, temperatures, storage volumes

Table 6.2. *Examples of some amenity and intrinsic values derived from water resource systems.*

Amenity and intrinsic values	Examples
Cultural, historical, heritage values	Existing historic buildings, bridges, waterworks, cultural landscapes
Geomorphologic values	Geomorphologic features such as sand banks, natural levees, sand dunes, point bars, natural levees, natural sedimentation processes, waterfalls
Landscape values	Beautiful views and landscapes
Religious values	Particular water bodies (such as in Israel and India)
Nature value	Rare species, existence of healthy ecosystems

One of the first steps in conducting environmental economic analyses is to determine the impacts on the environment and on other natural resources and ecosystems of the proposed water resource system project or policy in question. These impacts may be multiple, e.g., physical, biological, chemical, ecological or social. Project impacts are determined by comparing the with-project and the without-project scenarios.

An environmental impact can result in a measurable change in production and/or change in environmental quality. To estimate such impacts, one must usually rely on the expertise of engineers, ecologists, agronomists, social scientists and other experts. The impacts are multiple, complex, often poorly understood and cannot always be expressed in monetary units. However, a number of valuation techniques have been developed to help us compare environmental and economic impacts.

Methods for valuing the environment

In an attempt to place a monetary value on the environment, or on various environmental impacts that might result from a water resources or any other civil works project, economists refer to *valuing the environment*. In valuing the environment economists attempt to measure (1) individuals' preferences for environmental improvement or conservation, or (2) individuals' losses of well-being because of environmental degradation or from losing an environmental asset. They do this by developing and using revealed *willingness to pay* or *willingness to accept compensations* measures. They then make certain assumptions about one's ability to aggregate these individual valuations. There are many limitations to any of these approaches for valuing the environment (Pearce, 1993; Pethig, 1994).

While any economic benefit–cost analyses has its problems, so do alternative approaches for making social choice. Indeed, if water and what it supports in a natural environment have their own rights, not only those given to them by humans, then much economic development may be ethically wrong. Yet people have rights too. Since many decisions people make are based in large measure on economic benefits and costs, it is worth some effort to try to assign an economic or monetary value to the environment in order to give it more comparable status in the decision-making process.

While often preferable if possible, quantifying everything in monetary terms is not necessary in a multi-objective decision-making process. If different units of measure are used to quantify different objectives, various mulit-objective analyses (e.g., as in Steuer, 1986) can be used to help decide what mix of objective values is best in some sense. While much can be said for multi-objective decision-making, where some objectives or criteria are not all measured in monetary units, these methods eventually rest on subjective judgements. Those judgments are often more difficult to justify than are those made on the basis of money.

The basic concept of *economic valuation* is the individual's or society's willingness to pay for an environmental service or resource such as water. Valuation techniques based on actual behavior include change of economic productivity, loss of earnings, defensive expenditure, travel cost, wage differences, property values and artificial markets. Techniques based on potential behavior include replacement cost, shadow project and contingent valuation. Each of these are discussed in more detail below.

Under specific conditions, such as when the environmental impact leads to a change in the supply of a good or servide that is bought in a competitive market, the willingness to pay by current generations may be estimated directly using prevailing market prices. If the market is not fully competitive, then the market valuation will be only a partial measure. If the result of an impact cannot be directly related to a market activity, a lower bound on the willingness to pay may be able to be estimated by a closely related proxy such as loss of earnings, defensive expenditures, replacement costs and shadow projects. There may also be cases where the willingness to pay can be estimated through the derivation of the demand for the environmental asset based on actual behavior. Travel cost, wage differential and property valuation are possible measures. Finally, willingnes to pay may be elicited through controlled experiments or direct interviews, using artificial markets and contingent valuations.

These various valuing approaches will be discussed in some more detail shortly. Since all of them assume we can make some observations, none of them can be applied to future generations, unless we make the assumption that their willingness to pay will be similar to, or some function of, our willingness to pay. These are heroic assumptions.

Water resource projects can affect economic activities such as agricultural or industrial productivity positively or negatively. The incremental output can be valued by using standard economic prices. Projects can also have impacts on environmental quality and this in turn on human health. Ideally, the monetary value of health impacts should be determined by the individuals' willingness to pay for improved health. In practice, one may have to resort to second best techniques such as using foregone earnings in cases of premature death, sickness or absenteeism and increased medical expenditures. This approach may be relevant, for example, when considering projects that affect water pollution.

The *value-of-life approach* is often questioned on ethical grounds. Life is considered by some as having infinite value. But in practice, society implicitly places finite values on human life and health in policy and project decisions that affect environmental quality, safety and health. In the case of an increase or a reduction in the probability of numbers of deaths, an approximate estimate of value is the loss in estimated future earnings of the individuals involved.

Individuals, firms and governments undertake a variety of 'defensive expenditures' to avoid or reduce unwanted environmental effects. Environemtnal damages are often difficult to assess, but information on defensive expenditures may be available or can be obtained at lesser costs than direct valuations of the environmental good in question. Such actual expenditures indicate that individuals, firms or governments judge the resultant benefits to be greater than the costs. The defensive expenditures can then be interpreted as a minimum valuation of benefits. On the other hand, if these defensive expenditures are mandated by governments (and are not the result of market forces) these expenditures may not be valid measures of willingness to pay to reduce adverse environmental impacts.

Alternatively, replacement costs of a damaged asset can be used as a measure of benefits. But this may or may not exceed the damage of the asset. Hence this cost is only relevant if it would actually be incurred. The shadow project approach to estimating replacement costs involves the design and costing of an alternative project that would fully compensate those who are damaged or who incur economic and environmental losses due to a particular project. For example, if a proposed irrigation development project will reduce water allocated to current users, a shadow project would be one that replaces that water, or at least the value of that water, to those who will receive less because of the new irrigation development.

Valuations using implicit or surrogate markets use market information indirectly. These approaches use travel cost, property value, wage differrential and marketed goods as surrogates for non-marketed goods. Each technique has its particular advantages and disadvantages as well as its specific requirements for data and resources. the task of the analyst is to determine which of the techniques might be most applicable to a particular situation.

Travel cost is most often connected with recreational analyses in industrial countries. The *travel cost method* estimates the benefits produced by recreation sites based on the costs people are willing to pay to get there. A related method can also be used to value water from a water collection project, for example. In this case the area surrounding a site can be divided into concentric zones of increasing distance. A survey of users, conducted at the site, can determine the zone of origin, visitation rates, travel costs and various other socio-economic characteristics. Users close to the site would be expected to make more use of it, because its implicit price as measured by travel costs, is lower than that for more distant users. Analysis of the questionnaires should enable the construction of a demand curve based on the willingness to pay for entry to the site, costs of getting to the site and foregone earnings or opportunity cost of time spent. From this an associated consumers surplus (what they would be willing to pay in addition to what they actually pay) can be determined. This surplus represents an estimate of the net benefits of the water resource project, as applicable.

The *property value method* is based on the general land value approach. The objective is to determine the implicit prices of certain characteristics of properties. In the environmental area, for instance, the aim of the method is to place a value on the benefits of environmental quality improvements, or to estimate the costs of a deterioration due, for example, to pollution.

The *wage differential method* is based on the theory that in a competitive market the demand (wage paid) for labor equals the value of what is produced by that labor and that the supply of labor varies with working and living conditions in an area. A higher wage is therefore necessary to attract workers to locate in polluted areas or to undertake more risky occupations. Again, as in the case of the property value method, the wage differential method can only be used if the labor market is very competitive. This method relies on private valuation of health risks, not necessarily social ones. In this context, the level of information concerning occupational hazards must be high in order for private individuals to make meaningful tradeoffs between health risks and remuneration. The effects of all factors other than environment that might influence wages must be eliminated to isolate the impacts of the environment.

In cases where the water resource project provides non-marketed goods or services that have close substitutes that are marketed, the value of those goods or services can be approximated by the observed market price of its substitutes.

In the absence of people's preferences as revealed in markets, the *contingent valuation method* tries to obtain information on consumers' preferences by asking direct questions about their willingness to pay. The method involves asking people what they are willing to pay for some specified benefit (such as water one can swim in without health risks), or what they are willing to accept by way of compensation to tolerate a loss or damage (such as water they cannot swim in without risk). This process of asking may be either through a direct questionnaire survey, or by experimental techniques in which people respond to various stimuli in controlled conditions. What is sought are personal valuations of individuals for increases or decreases in the quantity or quality of some good, contingent upon a hypothetical market. Willingness to pay is constrained by the income level of those being questioned, whereas willingness to accept payment for a loss is not so constrained. Willingness-to-accept estimates therefore can be significantly higher than willingness-to-pay esitmates.

In certain circumstances, contingent valuation may be the only available technique for estimating benefits. It can be (and has been) applied to common property resources that everyone shares (e.g., air), amenity resources having scenic, ecological or other characteristics, and to other situations where market information is not available. Caution should be exercised in seeking to pursue some of the more abstract benefits of environmental assets such as existence value (of an asset that may never be used, but provides some kind of satisfaction to individuals merely because it exists).

Artificial markets could be constructed for experimental purposes to estimate consumer willingness to pay for a good or service. For example, water purification kits or accesses to a recreational area might be marketed at various price levels to estimate the value placed by individuals on water purity or on the use of a recreational facility.

When evaluating water resources projects that have negative environemtnal impacts, the shadow project approach can be used. This involves the design and costing of one or more alternative projects that provide for substitute environmental services to compensate for the loss of environmental assets under the ongoing projects. This approach is essentially an institutional judgment of the replacement cost, and is increasingly being considered as a possible way of operationalizing the concept of sustainability at project levels. It assumes that maintaining environmental capital intact is a constraint. Its application therefore could be most relevant when critical environmental assets are at risk.

MULTI-OBJECTIVE ANALYSES

The methods just described are sometimes used to estimate and express the environmental benefits and costs of a given water resoruces project in monetary terms. When projects and/or policies and their impacts are to be embedded in a system of broader objectives, some of what cannot be quantified in monetary terms, multi-objective analyses offer an alterative approach. To perform such analyses the objectives or criteria used to compare and evaluate alternative projects need to be specified. These objectives must be quantifiable, but not necessarily in the same units of measure, such as money.

Numerous multi-objective analyses can be used to identify dominated alternatives, at least with respect to the objectives identified and quantified (see, for example, Steuer, 1986). Dominated projects are those whose performance criteria, with respect to all evaluation criteria, are not superior to those of at least one other alternative project. Dominated alternatives can thus be removed from further consideration. If the removal or rejection of some apparently dominated alternative is resisted by some, then there must be some other, perhaps non-quantifiable, objectives being considered that should be explicitly identified. Experience indicates that this situation is commonly the case, and hence once again we express our view that these quantitative multi-objective analysis techniques are providing information, especially information on efficient tradeoffs among various objectives. They are not dictating optimal solutions.

There are numerous multi-objective modeling and planning methods. The best one to be used will depend on the project and on the preferences of those institutions and personnel involved in the project selection or decision-making process. The major feature of multi-objective models is that they permit the explicit consideration of multiple objectives expressed in different units. They force one to reveal preferences and consider tradeoffs between conflicting objectives. However, a key question concerns whose preferences are to be considered. While the modeling process helps individuals understand the issues and nature of the problem, the model solution simply provides tradeoff information to the decision-making process. Various interested groups may assign differrent priorities to the respective objectives, and in such cases it may not be possible to determine a single best solution from the use of any multi-objective model.

SOME ECONOMIC MODELS FOR CONSIDERING SUSTAINABILITY

This section explores some ways of incorporating economic objectives and constraints that pertain to sustainability within multi-objective water resource planning and management models. Two simple examples are offered, simply to illustrate two modelling approaches. The first example involves irrigated agricultural crop production. The issue being addressed in this example is what type of irrigation management should be selected considering the present generation of farmers and their descendants. The second example involves the withdrawal of non-renewable ground water for an aquifer. In the latter example, the withdrawal and consumption of this ground water is clearly not sustainable over a long period of time. Does that mean one should not use it in order to save it for future generations, and they in turn save it for their future generations? We will try to develop planning objectives for both situations that address sustainability issues, admittedly under a high degree of uncertainty, as will always be the case when anyone speculates about the distant future.

The approach proposed here for including sustainability in economic planning and management models is based on two observations. The first is that what we choose to do today is conditional on the resources, the knowledge, the environment, the institutions and all other things available to us today. We cannot go back into the past and change decisions that were made to give us a different set of conditions than those that we have today. However, we have it in our power to change the conditions facing our descendants. Hence, it would seem reasonable to at least guess at what our descendants would like us to do today to give them the conditions they would like to have as they carry on in creating an even better quality of life for themselves and, possibly, for their descendants too.

The second observation is that while we cannot predict with much confidence what future generations will want to do, or will care about and value, we can try to guess what they might want and value. We can then include these guesses, expressed as future objectives, along with our own present objectives in a multi-objective planning framework. If there are conflicts between what we who are living today want and what we think our descendants in future generations will want, then at least we can identify the tradeoffs. These are tradeoffs in the benefits we will derive from doing what we want to do for ourselves today and the benefits are descendants will get by doing what we think our descendants would like us to do for them. Such information should aid in any political debate on what actions or decisions are best.

Even though we cannot know what new objectives future generations will want to try to achieve, we can assume initially that their objectives will be the same as ours. If one of our objectives is to maximize the present value of net income, we can assume this objective will be among the objectives our descendants will want to achieve as well. In this simple illustrative example we will assume that an income maximization objective is the primary objective of both generations (ours and our descendants), although we know the real situation certainly is and will be much more complex.

To introduce the general modeling approach, assume the conditions or resources available to us who live today are denoted by a vector, $\mathbf{R}_0$. Also assume our overall objective is to make decisions, $\mathbf{X}_0$, that will maximize the present value of our total welfare, $W_0(\mathbf{X}_0|\mathbf{R}_0)$, however measured, given the conditions (resources), $\mathbf{R}_0$, we currently have. The vector of decisions we make, $\mathbf{X}_0$, will impact or determine, in part, not only our own welfare but also the conditions or resources, $\mathbf{R}_y$, available in the future periods (years), y. Like us, our descendants in future periods, y, may want to maximize the present value of their welfare, $W_y(\mathbf{X}_y|\mathbf{R}_y)$, given their current conditions, $\mathbf{R}_y$. Their current conditions, $\mathbf{R}_y$, are dependent on the resource management decisions of those who preceded them.

Since we who are responsible for planning and managing our resources today have some impact on the conditions facing those of future generations, we have an obligation to at least consider not only our own immediate objectives but also those desires or objectives of future generations. This can be done in a multiple-objective framework. Letting a_y be a relative weight given to a future welfare function, $W_y(\mathbf{X}_y|\mathbf{R}_y)$, in period y, a multiple-objective function can be defined that includes the welfare of a series of periods beginning now and terminating at some time in the future:

$$\text{Maximize } \Sigma_y \, a_y W_y(\mathbf{X}_y|\mathbf{R}_y) \qquad (6.1)$$

where each set of future conditions, $\mathbf{R}_y$, depends on the current conditions, $\mathbf{R}_0$, and on the sequence of decisions, $\mathbf{X}_0, \mathbf{X}_1, \mathbf{X}_2, \mathbf{X}_3, \ldots, \mathbf{X}_{y-2}$, and $\mathbf{X}_{y-1}$, up to period y.

This multiple-objective modeling approach can be used to determine tradeoffs among various future generations by varying the values of the relative weights, a_y. More importantly, however, this modeling approach can help us identify today what future generations would like us to do for them. It should be noted, however, that whether or not we choose to do for them what we think they would want us to do is a political decision, an issue given more attention later in this chapter.

Irrigation planning

There are a number of options for growing crops. Each involves different management practices, and different amounts of inputs such as water, fertilizer, pesticides, labor, machinery, etc. For the purposes of this simple example, consider a situation in which there is a choice between two irrigation management practices for the production of a single crop on a single soil type. One management practice will be called intensive (i.e., high water use) and the other will be called extensive (relatively less water use).

The intensive irrigation management practice will require considerable irrigation and drainage infrastructure and water. It will also result in higher crop yields initially, but because of predictable increases in soil salinity, these high initial crop yields will diminish over time.

The extensive irrigation management practice will require less infrastructure and water and will not result in increased soil salinity concentrations. Hence little if any crop productivity degradation will result, but current crop yields per hectare will be substantially less.

Clearly the extensive irrigation management practice is the more sustainable one of the two management options if sustainability is measured in terms of long-term non-decreasing crop yields. It may turn out that future generations won't care, but at least the option is open to them if we today choose the extensive option. The purpose of this example, however, is not to prove this obvious fact. It is to show how traditional economic efficiency maximization objectives (for both present and future generations) can be included in a planning model, and how tradeoffs between the profits of current and future generations may be identified.

The example could also involve the use of ground water to supplement surface water in periods of low surface water supplies. If the intensive management practice were to require withdrawals in excess of recharge rates over time, clearly the aquifer water table (assuming a water table condition) would lower and the cost of pumping would increase. If pumping costs do not become too high, the aquifer could dry up (although this is seldom the case). In any event, crop production will become more expensive for those farmers dependent on ground water. Eventually they may be unable to compete with farmers not dependent on these diminishing groundwater supplies. As farming that depends on diminishing groundwater supplies becomes more marginal, management practices often become more short-sighted, resulting in additional long-term negative impacts, such as soil degradation.

Crop production

In this single-crop, two irrigation management practice example, assume that each management practice, denoted by the index m, results in a decreasing crop yield per hectare of $Y(y, m)$, in successive periods, y, on into some distant future. This decrease in crop yield is dependent on the management practice, m. The irrigation management index, m, can assume two values: intensive ($m = i$) and extensive ($m = e$) management. This in each period, y

$$Y(y + 1, m) \leq Y(y, m) \tag{6.2}$$

where the initial (current) per hectare yields are $Y(0, m)$ and a strict inequality applies for $m = i$, the intensive management practice. These crop production functions, $Y(y, m)$, assume specific allocations of the required resources necessary for crop growth and harvest, on a single soil type.

Assume that the annual per hectare decrease in yield associated with intensive management is greater than for the extensive management practice. Also assume the initial (current) crop yield per hectare from intensive management, $Y(0, i)$, is higher than the more sustainable extensive per hectare yield, $Y(0, e)$.

Resource limitations

In this example land is the limiting resource for both irrigation management options. Part of the land can be managed intensively and part managed extensively. The purpose of the model being constructed is to find out how much of the land is to be managed under intensive and extensive management. Put another way, how much land is to be sustained, and how much is to be degraded in the pursuite of increased crop yields.

Obviously the total hectares under irrigated agriculture crop production need not equal the total land available. However, in this income-maximizing example, where profits can be made under either management practice and since land is the limiting resource, the total number of hectares under management will equal the total hectares available. Because of the substantial investment costs as well as the fact that income obtained from selling the crops is to be maximized, assume that the land that is to be managed intensively will remain under intensive management. Similarly, for land to be managed extensively. Our task is to determine the hectares under both types of irrigation management.

Denoting $L(i)$ and $L(e)$ as the unknown amounts of land to be devoted to intensive and extensive irrigation, the total land constraint is

$$L(i) + L(e) \leq \text{Maximum land available} \tag{6.3}$$

Other assumptions could have been made. If water were the limiting resource and intensive agriculture required more water per hectare than extensive agriculture, obviously more land could be allocated to extensive agriculture than would be possible for intensive agriculture given the limited amount of water available. Depending on the cost per hectare of crop production, the increased acreage of extensive production could result in higher net income than would be possible using intensive management on less land.

Total crop production

Equations defining the total crop production, $TY(y, m)$, in each period, y, and for each management alterantive ($m = i, e$), are simply the yields per hectare, $Y(y, m)$, times the number of hectares under management, $L(m)$:

$$TY(y, i) = Y(y, i) \cdot L(i) \tag{6.4}$$
$$TY(y, e) = Y(y, e) \cdot L(e) \tag{6.5}$$

Income and costs

Assume that the end-of-period price per unit crop, $P(y)$, will increase at a constant annual rate of inflation, inf. Similarly for the production costs per hectare, $C(m, y)$, which are dependent on the management practices, m. Further, assume the costs must be paid at the beginning of the period, and the income will be received at the end of the period. Denote the annual compound interest rate, that includes the rate of inflation and the real rate (time value of money), as r. Thus the net income, $NI(y)$, from agriculture crop production in at the end of period y is:

$$NI(y) = \Sigma_m[P(y) \cdot TY(y, m) - C(m, y) \cdot L(m) \cdot (1 + r)] \tag{6.6}$$

where the sum is over m equal to i and e, and because of inflation:

$$P(y) = P(0) \cdot (1 + inf)^y \tag{6.7}$$
$$C(m, y) = C(m, 0) \cdot (1 + inf)^{y-1} \tag{6.8}$$

The present value (at the beginning of any current period, $y = 1$, which is the end of period 0) of the total net income for all future periods, y, to some terminal time, Y, is:

$$PVI(1) = \sum_{y=1}^{Y} NI(y)/(1 + r)^{y+1} \tag{6.9}$$

Management objectives

For farmers interested in being economically efficient, their objective would be to maximize the present value of their total net income beginning with the current year, $y = 1$:

$$\text{Maximize } PVI(1) \tag{6.10}$$

This is a reasonable objective no matter which year is being considered, yet the desired management practice results may be quite different depending on previous irrigation management practices.

Clearly, farmers today can't change what has already happened to their land in the past. However, they can consider what future generations might have wished they had done, now, as they select their current resource management practices.

Assuming a planning horizon together with a variety of prices, costs and discount rates, the solution of the above model will show the obvious. Intensive management on all the available land will maximize the present value of net income. But what of the future?

Considering the future

There are several options for considering the future. One option is to alter the discount rate used for planning, the compound interest rate, r, applicable for the period y. As the discount rate, r, is lowered, the present value of net income, $PVI(y)$, contains increasing amounts of the future net incomes, $NI(y)$. Conversely, if the discount rate, r, is increased, what happens in the future becomes less important at the present.

There is some debate among economists over the appropriateness of lowering discount rates to give more weight to the future. While future net benefits are given increased importance with lower discounts rates, r, more marginal projects, ones that may or may not be good for the environment or result in sustainable policies, also become more economically attractive. *Thus, it is not obvious that lower discount rates will always result in more sustainable resource management policies.* In fact, just the opposite is likely. Decreasing the discount rate, r, is used in this example problem does not always make extensive (sustainable) management more attractive than intensive management.

An alternative approach for considering the futurue is, as mentioned above, to place ourselves in our descendants shoes and assume that their objectives will be the same as our objectives today, e.g., in this example the maximization of the present value of net income from farming. The difference is that this future objective can be conditional on what decisions we make at the present time. The purpose of this is to determine what mix of management practices future generations would like us to implement today. This does not mean we must or will necessarily do what our descendants wish, only that we want to identify what those wishes might be. We can then examine the range of tradeoffs, if any, between what we want to do and what we think those who follow us in the future might want us to do.

To do this in this irrigation example we must define the objective functions of future farmers. Assume for this example that economic efficiency (i.e., net income maximization) will continue to be the objective of future generations. Given a constant annual discount rate, r, which includes the inflation rate, the present value of net benefits in period y based on the following Y periods is:

$$PVI(y) = \sum_{n=y}^{Y+y} NI(n)/(1+r)^{n-y+1} \tag{6.11}$$
$$\text{for } y = 1, 2, 3, \ldots Y$$

The multiple objective function that includes the present values of the next N periods can be defined as the sum of weighted successive future present values:

$$\text{Maximize} \sum_{n=1}^{N} a_n \cdot PVI(n) \tag{6.12}$$

The relative weights, a_n, could vary from 0 to 1 and their sum could equal 1. Clearly, if a_1 is 1 and all the other weights are 0, this is the same objective as Eq. 6.10, above.

Using the present values defined by Eq. 6.11 one could maximize objective Eq. 6.10 or Eq. 6.12, subject to constraints specifying that the successive present values, $PVI(y)$, for increasing future periods, y, do not decrease. This enforces the economic surrogate of sustainability.

$$PVI(y+1) \geq PVI(y) \tag{6.13}$$
$$\text{for all periods } y = 1 \text{ to } y = Y - 1$$

Some results

Using the above example and varying the weights, a_n, in Eq. 6.12; the interest rate, r, the costs, $C(e, y)$ and $C(i, y)$, and price per unit crop, $P(y)$, the yields, $Y(y, m)$, associated with extensive and intensive management (with the inclusion or exclusion of constraints (Eq. 6.13) above) the following general conclusions can be made:

- Decreasing the discount rate will not always result in the selection of a sustainable policy.
- Imposing constraints (Eq. 6.13) ensuring non-decreasing present values may not (in this example, at least) ensure fully sutainable management practices given certain values of the discount rate, price and costs.
- Maximizing the present value today of future net income will not result, in this example, in a sustainable policy.

- As the weights assigned to future objectives (in Eq. 6.12, above) increase relative to those assigned to present objectives, the current optimal management policy, in this example, shifts from one that is non-sustainable to one that is sustainable. This occurs as the total net present value of income at the current time decreases. This is the tradeoff that must be considered in this admittedly simple and limited example for deciding how many hectares to irrigate intensively and/or extensively.

Groundwater mining

Ground water can be used beneficially to augment surface water supplies in times of surface water shortages. They can then be recharged during times of surplus surface waters. Sustainable use of groundwater resources normally implies drawing no more from the aquifers than their average recharge rates. If the long-term average withdrawal does not exceed the long-term average recharge, the conjunctive surface/groundwater supply system can provide a higher sustained yield than that available only from surface water supplies. The maximum duration and extent of the dry and wet periods which can be sustained will depend, in part, on the aquifer storage capacity.

Two conditions can cause the gradual, or not so gradual, long-term depletion of groundwater supplies. One condition is when the growth in demand for water begins to exceed the long-term average surface water supply. When this happens ground water can then be used to make up the ever increasing deficit. The other condition is when the amount of ground water in storage is being used to meet the demand on a regular continuing basis and the demand exceeds the recharge rate.

Arid regions deficient in surface water supplies but having some groundwater supplies that are not being replenished, are faced with a challenge. How should those non-renewable groundwater supplies be used in a manner that promotes as much as possible the long-term economic and social development of those regions? How much of this non-renewable groundwater resource should be pumped and used, and when, in order to enhance the overall level of sustainable development?

Many involved in renewable resource management tend to equate sustainability with resource preservation. This cannot apply to non-renewable resources, including ground waters in aquifers having relatively low natural recharge. If it did it would require preserving those groundwater resources so that future generations would have the same opportunities for their use as we have today. Those future generations

would have to preserve it for their future generations, and so on into the distant future. Such a policy is certainly sustainable with respect to resource preservation, but it does not offer any advantage to any of the generations who might benefit from the use of some of that resource. *A policy focused on resource preservation would not likely be in the best interests of the region or country.*

Clearly, if a preservation policy were implemented today, it would severely limit the economic and social development of arid regions that are currently depending on these groundwater resources for their continued development and welfare. Would future generations want to be born into an economic and social environment much the same or perhaps worse than that which exists today, but having available more groundwater supplies? Or would they want today's generation to continue to develop their economic and social systems at the cost of using up a little (or even a lot) of today's essentially non-renewable ground water? We will never know for sure. But decisions are being made today with respect to the 'mining' of non-renewable groundwater supplies and these decisions will certainly impact future generations.

It is possible that by abstracting and consuming ground water in this generation a region or country could become better developed economically and socially and become much more capable of developing and using new technology to enable it to obtain substitute sources of water at reasonable costs in future generations. Does it make much sense, then, to save or preserve today's resources, including ground water, needed to make this happen? Yet no one can be sure that this hoped for improvement in technology and increased knowledge would actually happen. So, what is the appropriate allocation of non-renewable resources over time? Clearly any analysis of economic and social sustainability issues will require planners to look at more than just the water sector, as critical as that sector is for development.

The central question is just how much of a non-renewable groundwater resource should be used now and in the future, recognizing that the future is important and it is also very uncertain. We cannot easily use the multi-objective approach of the previous irrigation example, because we do not know the long-term economic and social impacts of using any of the groundwater resources today. We do know, however, that what we choose to use now of this non-renewable resource will not be available for use in the future. It is certainly possible that the economic and social benefits to future generations could be greater if we use some of that resource today rather than saving it to be used in the future. On the other hand, those future benefits could be less. In any event, the future benefits and losses are not easily quantified.

Hence, a set of single-objective models is sketched out below. These might be updated and used periodically, on into the future, to guide the adaptive use and allocation of non-renewable groundwater supplies over space and time. The set of models and their solutions identifies some economic relationships that should influence, but not necessarily dictate, the use of limited non-renewable resources over time.

This example, like the previous one, merely illustrates a modeling approach. The objective in each period on into the future is assumed to remain one of economic efficiency. We are well aware that there are other important objectives, and no doubt other constraints as well. Any actual application of this illustrative modeling approach will undoubtedly require model modifications and expansions. Only then can they continue to be useful to those responsible for determining any adaptable groundwater pumping schedule in regions overlying non-renewable groundwater supplies. Here the model is kept simple on purpose.

Allocations over space in a single time period

To begin, consider first a situation in which there is an existing groundwater supply, W, no recharge, a number of water users ($u = 1, 2, \ldots, U$), and only one time period available to use the water. This is a simple resource allocation problem, in which the allocations, x_u, to each use, u are to be determined so that the total allocation does not exceed the total groundwater resource available, W. Assuming the objective to be satisfied is the maximization of net economic benefits (the sum of known benefit functions, $B_u(x_u)$, less pumping cost functions, $C_u(X_u)$, the model can be written:

$$\text{Maximize: } \Sigma_u[B_u(x_u) - C_u(x_u)] \tag{6.14}$$
$$\text{Subject to: } \Sigma_u x_u \leq W \tag{6.15}$$

Optimal values of each allocation, x_u can be determined using Lagrangian multipliers. Assuming the demand for water exceeds the amount available in the period so that the resource constraint above becomes an equality, the Lagrangian function, $\mathbf{L}(\mathbf{x}, \lambda)$, of the vector of unknown variables, x_u, and the single unknown constraint multiplier λ, is:

$$\mathbf{L}(\mathbf{x}, \lambda) = \Sigma_u[B_u(x_u) - C_u(x_u)] - \lambda \cdot (\Sigma_u x_u - W) \tag{6.16}$$

Optimality (economic efficiency) conditions can be derived by equating to 0 the partial derivatives of Eq. 6.16 with respect to each of the unknown variables, x_u and λ. These conditions require that the marginal (change in) net income derived from the last unit of water allocated to each use be equal for all users. Furthermore this marginal net income should equal the value of the Lagrange multiplier, λ. This multiplier, λ, is the marginal change in total income to all uses, u, with respect to a change in the total water available, W. Economic efficiency also requires that when the total demand exceeds the total supply, the total allocation to all uses equals the amount of water available, W. Since demand is assumed to exceed supply, no water should be wasted. If these optimality (economic efficiency) conditions are not met, the total income that can be derived from the groundwater supply, W, is not being maximized.

Total benefit and cost functions for each period

Note that whatever the amount of a resource to be allocated in each period of a multiple period problem, the same optimality conditions will apply in each period. For economic efficiency, all uses should have identical marginal net benefits. Using this fact we can construct a total benefit function, $B_y(X_y)$, and a total pumping cost function, $C_y(X_y)$, associated with a total amount of ground water, X_y, pumped and allocated in period, y. Using these total benefit and total pumping cost functions, the above model, written for a specific period, y, can be expressed as:

$$\text{Maximize: } B_y(X_y) - C_y(X_y) \tag{6.17}$$
$$\text{Subject to: } X_y \leq W_y \tag{6.18}$$

the resource available at the beginning of period y. Assuming, of course, that the most economically efficient allocations, $x_{u,y}$, are made to each use, u, in each period, y

$$\Sigma_u x_{u,y} = X_y \tag{6.19}$$

Clearly the optimality conditions that apply over space (i.e., to different users in the same time period) also apply over time. The marginal net benefits derived from the last portion of the total allocation, X_y, will equal the marginal net benefits associated with a marginal addition to W_y.

Multiple time periods models

To extend this model to multiple periods, Y, define the functions $PB_y(X_y)$ and $PC_y(X_y)$ as the present value of the total benefits and pumping costs assoicated with an amount X_y of water in period y.

$$PB_y(X_y) = B_y(X_y)/(1 + r)^y \tag{6.20}$$
$$PC_y(X_y) = C_y(X_y)/(1 + r)^y \tag{6.21}$$

where r is the period discount rate.

Again, assuming no recharge, the model for determining the amounts, X_y, to be pumped from the groundwater aquifer each period, y, given a supply of W, becomes:

$$\text{Maximize: } \Sigma_y[PB_y(X_y) - PC_y(X_y)] \qquad (6.22)$$
$$\text{Subject to: } \Sigma_y X_y \leq W \qquad (6.23)$$

This model has exactly the same form as the single-period model (Eqs. 6.14 and 6.15), and hence the conditions for optimality are the same. The marginal present values of net benefits must be the same for all periods, y. They must all equal the marginal present value of net benefits associated with a marginal addition in the total available resource, W.

In the above multi-period model the single resource constraint could be written as a series of continuity constraints, one for each period y.

$$W_{y+1} = W_y - x_y \text{ for all periods, } y \qquad (6.24)$$

Again, W_y is the amount of resource available in the beginning of period y, and W_0 is the known total amount currently available. In this case the conditions for optimality derived from equating to 0 the partial derivatives of the Lagrangian

$$\mathbf{L}(\mathbf{X}, \mathbf{W}, \lambda) = \Sigma_y[PB_y(X_y) - PC_y(X_y)] \\ - \Sigma_y \lambda_y \cdot (W_{y+1} - W_y + X_y) \qquad (6.25)$$

require the same conditions for economic efficiency as the previous model. The marginal present values of net benefits in each year should be equal, and each of them should be equal to the marginal present value of the total net benefits in each year. Each multiplier, λ, in each period, y, should be equal.

Given the above conditions required for economic efficiency, we can draw some conclusions regarding the relative amounts of water to be pumped from the ground in successive periods, y. If the marginal present values of the net benefits associated with any amount pumped are increasing over time (perhaps due to an increasing water scarcity or to increasing water use efficiencies and values) the economically efficient allocations, X_y, will also have to increase over time. On the other hand, if discounting or increasing costs create decreasing marginal net benefits associated with any quantity of water, X_y, over time, the economically efficient amounts of water to be pumped and allocated should also decrease over time. Otherwise the marginal present values of each period's net benefits will not remain the same.

If one assumes that every period's net total net benefit function, $B_y(X_y) - C_y(X_y)$, is the same, then the allocations, X_y, will decrease by a fixed percentage. That percentage will equal the discount rate, the rate of interest used to convert each future net benefit function to its present value.

Considering recharge and imports

One reason for replacing the single resource constraint with a resource continuity constraint (Eq. 6.23) for each period, y, is that we now can easily introduce any estimate of the natural groundwater recharge, R_y, in each period, y.

$$\text{Maximise: } \Sigma_y[PB_y(X_y) - PC_y(X_y)] \qquad (6.26)$$
$$\text{Subject to: } W_{y+1} = W_y + R_y - X_y \qquad (6.27)$$

for all periods, y.

The water available in period y is now $W_y + R_y$. The addition of an estimated non-zero natural recharge may change the marginal present values of net benefits, but not the optimal relationships among those marginal net benefits. These marginal present values of net benefits should still be equal over time. Equation 6.27 instead of Eq. 6.24 applies in each period, y.

Next consider the possibility of importing an unknown amount of water, I_y, in each period, y, at a present value cost of $PIC_y(I_y)$. This additional consideration added to our groundwater withdrawal model converts it to a multi-source model. Multi-source models apply to many water users having access to both surface and groundwater supplies. Adding these import and cost terms to the above model yields:

$$\text{Maximize: } \Sigma_y[PB_y(X_y + I_y) - PC_y(X_y) - PIC_y(I_y)] \qquad (6.28)$$
$$\text{Subject to: } W_{y+1} = W_y + R_y - X_y \qquad (6.29)$$

for all periods, y.

This introduces an additional set of conditions for economic efficiency, namely that the marginal net benefits derived from the last unit of imported water must equal the marginal cost or providing that last unit of imported water. This is a logical conclusion. If imported water costs more than it is worth, by definition it is not worth importing and benefits are being lost.

Increasing pumping costs with depth of ground water

Finally, consider the pumping cost not only as a function of the amount of water pumped, X_y, but also as a function of the existing groundwater storage, W_y in each period, y. A lower storage can result in a greater pumping head, and hence in a higher pumping cost. Given this assumption, the pumping cost term becomes $PC_y(X_y, W_y)$. The conditions for optimality (economic efficiency) now require that the marginal cost of pumping associated with a change in W_y must equal the difference in the multipliers, $\lambda_\psi - \lambda_{\psi-1}$. This difference is the change in the marginal present value of total net benefits from the last period to this period.

Until this new pumping cost component was introduced, the marginal present values of the total net benefits in each period were to be the same. The maximum net benefit derived from the use of the ground water could be obtained by equating the marginal present values of total net benefits in each period, y. Now these marginal present values of total net benefits must differ by the marginal present value of the cost of pumping. The marginal present value of the net benefit associated with an allocation of X_{y-1} in period $y-1$, plus the change in the marginal present value of the pumping cost due to a change in W_y. This increasing marginal present value of net benefits over time associated with the optimal values of allocations, X_y, will cause a decrease in the successive values of X_y, as one would expect.

As before, the marginal present value of the net benefit associated with an allocation X_y in period y, which is now changing over time, must equal the marginal present value of the net benefits associated with a change in the available resource, W_y, in period y.

Some practical considerations

We have proposed models for different conditions and have obtained some general guidelines for the pumping and use of a non-renewable groundwater resource over time. It is reasonable to ask how these models and their results can be used to guide water resource planners and managers in practice. In practice not all water use benefits can be expressed in monetary terms. Furthermore, economic efficiency is not the only objective to be met. Even if it were, we do not know future costs, benefits and interest rates. At best, we can only guess what they might be. We can only guess at current and future groundwater recharge rates, and in fact we may not even know for certain the amount of our currently available groundwater resources. And we can only imagine some kind of future technology that might allow future generations other options that may reduce their demand for non-renewable groundwater supplies.

However, ignorance about the future is not an excuse for inaction or myopic (nearsighted) decision-making. While all we can do today is to guess about the future, we must make the best estimates we can of future conditions, or possible scenarios of future conditions, and based on current and estimated future conditions, make our decisions about how much to withdraw now. Future decisions can be based on updated information on then-current and future conditions. While what is important now is only this period's decision, it

should be made considering the best information we have and the best assumptions we can make about what conditions (interest rates, costs, benefits, etc.) will exist in the future.

The time horizon of the model should be long enough so that the computed allocations to water users in the current and next few periods are not affected by some arbitrary terminal time horizon. This approach could involve modeling successively larger periods and aggregating the objective functions and recharge and pumping quantities for the years within those periods. Alternatively, one could set a limit to the total amount of ground water to be used over specified durations in the future.

Based in part on the information derived from the solutions of such models, decisions could be made with respect to pumping in the current and perhaps next several periods. Only that part of the solution defining the policy for the next several periods is of interest. Before the end of those few periods, the same modeling procedure could be carried out again, after updating the model's data, to determine the following several periods pumping schedule.

The information derived from this adaptive and sequential modeling process could suggest some modification in the pumping and allocation policies from one decision period to the next. One would not expect major changes unless the updated data, or the objectives, or the relative weights assigned to multiple objectives were significantly different from those used earlier.

SOME CONCLUDING COMMENTS ON ECONOMIC MODELING

Some simple examples have illustrated how multi-period economic efficiency objectives might be defined and included in a multi-objective modeling framework for considering sustainability in the planning and management of water resources systems. Economic efficiency objectives are not the only ones that should be considered. One can only guess at what future generations will select for their objectives, but among them might be objectives similar to those we use today.

After attempting to define the economic and other objectives of future generations, we can use them in models for estimating the tradeoffs between what we would like to do today for ourselves and what our descendants might like us to do today for them tomorrow.

Since we cannot look into the future with much precision, and since our models require us to make guesses about the future, this modeling process should be a sequential and adaptive one. The assumptions concerning the future should be re-evaluated each time planning or decision-making takes place. Our estimates of future desires, technology and economic and environmental conditions will always need updating based on the most recent information we can use to guess about the future. All we can do is to make our best estimates at the time we must make these estimates. Then, based on these estimates, generate information useful to planners and decision-makers and allow the political process to determine the appropriate tradeoffs that must be made between present and future generations.

While models can help identify and estimate tradeoffs among multiple objectives of present and future generations, it is not a trivial matter to introduce this information and get it used in any decision-making process. The actions of interest groups or lobbies are often what determine the relative importance placed on various multiple objectives. There are few if any lobbies representing the interests of future generations. There are many representing the interests of present generations. It is difficult to see a way to change this situation except through continued education, debate and public discussion of sustainability issues.

SOME THINGS TO REMEMBER ⎯⎯⎯⎯⎯⎯

- An inventory of all potential beneficial uses that can be provided by a particular water resource system is one of the first steps in any planning process.
- The identification of sustainable development options requires, among other factors, an estimate of the economic value of any damage to the environment that any water reosurces system design or operating policy may cause. This is essential information for making environmentally sound investment decisions – but not always easy to obtain.
- Environmental economics plays a key role in identifying options for efficient water resource management. It provides a bridge between the traditional techniques and the emerging more environmentally sensitive approach to decision-making. It will help to incorporate ecological concerns into the conventional economic decision-making framework.
- While often preferable if possible, quantifying everything in monetary terms is not necessary in a multi-objective decision-making process.

- Valuing approaches assume we can make some observations, none of which can be applied to future generations, unless we make the assumption that their willingness to pay will be similar to, or some function of, our willingness to pay.
- When projects and/or policies and their impacts are to be embedded in a system of broader objectives, some of which cannot be quantified in monetary terms, multi-objective analyses offer an alternative approach.
- There are numerous multi-objective modeling and planning methods. The best one to be used will depend on the project and on the preferences of those institutions and personnel involved in the project selection or decision-making process.
- While we cannot change the past, we do have it in our power to change the conditions facing our descendants. It would seem reasonable, then, to at least guess at what our descendants would like us to do today to give them the conditions they would like to have as they carry on in creating an even better quality of life for themselves.
- We can sometimes assume initially that the objectives of future generations will be the same as ours, although we know that the real situation is and will be much more complex.
- If the long-term average withdrawal does not exceed the long-term average recharge, the conjuctive surface/groundwater supply system can provide a higher sustained yield than that available only from surface water supplies.
- It is not possible to equate sustainability with resource preservation of non-renewable resources.
- The central question is just how much of a non-renewable resource should be used now and in the future, recognizing that the future is important and it is also very uncertain.
- Ignorance about the future is not an excuse for inaction or myopic decision-making. While all we can do today is to guess about the future, we must make the best estimates we can of future conditions, or possible scenarios of future conditions.
- Since we cannot look into the future with much precision, and since our models require us to make guesses about the future, the modeling process should be a sequential and adaptive one, the assumptions concerning the future being re-evaluated each time planning or decision-making takes place.

7 Ecological and environmental sustainability criteria

ECOSYSTEM AND ENVIRONMENTAL IMPACTS OF WATER RESOURCE SYSTEMS

Natural and managed water resources systems serve many purposes, and provide multiple benefits to individuals using those resources. Many of the benefits can be expressed in monetary units. Others cannot. Among the benefits and costs that are at best very difficult to express in economic terms are those derived from the environment and ecosystems that are associated with and that are dependent on those water resource systems. Many of those benefits that can be expressed in monetary terms are strongly dependent on the quality of the environment and on the health of the natural ecosystems associated with the water resource systems.

The protection and maintenance of healthy natural ecosystems and environments can also be strongly dependent on the vitality of the economy. One need only look at the destruction of natural environments and ecosystems in economically depressed regions to see this. Those living in these regions will surely become poorer as their economic and life support systems are diminished. The degradation of environments and ecosystems will not only result in long-term economic costs, but also in declines in the health and vitality of most living plants and animals. Humans will be affected directly and indirectly by the quality of the environment in which they live.

Benefit/cost analyses of economic development alternatives are incomplete unless the value of environmental and ecosystem impacts are included in those analyses. However, because it is difficult to express environmental and ecosystem benefits and costs in monetary terms, such comprehensive benefit/cost analyses are not easily per-

formed. Nonetheless, valuing the environmental and ecological impacts of alternative water resource system designs and operation policies, while difficult, is essential for the sustainable long-term planning, development and management of these systems.

Clean water is a necessary input for many economic activities. While the value of the clean water as an input to economic activities can often be expressed in monetary terms, any resulting damages or costs of waste residuals deposited in water bodies from these economic activities are not as easily valued. Indeed, one of the traditional functions or benefits of surface water resource systems has been the transport of such wastes away from the point of discharge, with the assimilation of at least a portion of these wastes by natural biochemical and physical processes as they travel downstream. But the economic value of such in-stream transport and waste assimilation is not just the cost savings of waste treatment and disposal at the discharge sites. If combined waste discharges are excessive, for example, they can damage the assimilative capacities of the water resource systems and their associated ecosystems. The resulting degradation could create losses for a number of other in-stream and off-stream uses of the water resource.

Droughts, floods, and erosion can also cause damage to humans and their property. These relatively rare (or 'extreme') events are merely the result of the natural variability of the supply of water in comparison to what is normally expected or observed. To the extent that humans have made themselves dependent on non-extreme conditions, and have not provided for or have not protected themselves against these occurrences, such events can cause damage and even death. The damages and losses of lives result because of conflicts between natural and socio-economic systems. Similar conflicts exist whenever toxic wastes are discharged into water bodies used for water supplies, the maintenance of aquatic ecosystems or for recreation.

To help understand and predict the time-varying environmental and ecological processes and patterns in natural aquatic ecosystems there are some parameters that can be measured. Among the more important ones are the time patterns of water flows, depths, and velocities that together influence erosion and sedimentation processes, the frequency and extent of wetland or floodplain inundation, the quality of the water and the type of soil and riparian vegetation along the water courses. These characteristics of water resource systems and their associated ecosystems will vary over time, and also over space. They will vary within and among river basins, depending on the geologic and geomorphologic character of the river beds, the climate of the region, and the quantity and quality of the water as impacted by human activities.

Aquatic ecosystems may also depend on continuity. For example, many fish and other aquatic species depend on the continuity and integrity of the aquatic ecosystems throughout a river system. Continuity of an acceptable aquatic ecosystem along a water course is crucial for salmon in many rivers like the Columbia River in northwestern USA, the Fraser River in Canada and the Rhine River in western Europe. Salmon survival depends on the continuity of a favorable habitat throughout the river if they are to enter the mouth of those rivers and travel great distances to their spawning grounds.

Some of the many environmental and ecological benefits and functions served by water resource systems and their watersheds are listed in Table 7.1.

Each of the functions listed in Table 7.1 are influenced by the pattern or time series of flows, velocities, qualities, soil characteristics and the like. Estimates of the time series in the future can be obtained from a water resource system simulation model. They will depend in part on the development and management policy assumed to be in effect in the future. To assess which time-series patterns (and thus which applicable development and management policies) are satisfactory and which are not with respect to each of the functions water resource systems can serve is often a challenge. Ecosystem enhancement (and even fish survival) is not always simply a function of some minimum level or range of in-stream flows, velocities and qualities. Rather it is the entire regime that may be important. Hence flow, velocity and quality time-series data for entire water resource systems over relevant periods must be assessed with respect to their impacts on each of the important environmental and ecological criteria.

Table 7.1. *Some environmental and ecological functions of natural watersheds*

Water resouce, ecosystem and wetland functions	
Natural flood and erosion control	Reduce velocities, flood peaks, wind and wave impacts, soil erosion
Water quality enhancement	Reduce sediment loads, filter for nutrients and impurities, process organic and chemical wastes, moderate temperature of water
Maintain groundwater supply	Promote infiltration and aquifer recharge capability, reduce frequency and and duration of low flows, increase reliability of conjunctive surface/groundwater supply systems
Water supply	Water for irrigation, municipal, industrial and energy uses
Navigation and recreation	Efficient transport of bulk cargo, inland waterways for recreation boating
Flora and fauna enhancement	Maintain biological productivity, maintain breeding and feeding habitats, create and enhance fish and waterfowl habitats, protect habitats of rare and endangered species
Cultural, scientific and other impacts	Maintain and enhance agricultural lands, forests and other silvicultural resources, recreational and open space activities, aesthetic values, scientific research opportunities, archeological and historic sites

Using any of the increasing number of available ecosystem simulation models that can predict ecological as well as environmental variables, time series can be generated. Each of these time series can be divided into successive sub-periods, averaged within each sub-period to remove natural variations within short periods of time, and then analyzed with respect to trends – trends leading toward increasing or decreasing sustainability (as discussed in greater detail in Chapter 4.) The analyses are clearly dependent on the many assumptions incorporated into the predictive simulation models and on the subjective judgments as to which environmental or ecological states are satisfactory and which are not, and which water resource regimes or conditions lead to such satisfactory states and which do not. If the trend over time shows an increasing frequency of 'satisfactory' sates, the system can be considered as becoming increasingly sustainable – under the assumption of the criterion or set of criteria assumed.

AN ADAPTIVE APPROACH TO WATER RESOURCE SYSTEM PLANNING AND MANAGEMENT

Closely related to the sustainability concept is the so-called adaptive approach to ecosystem and water resources planning and management. This is a goal-driven approach to restoring and sustaining healthy natural resource ecosystems and their functions and values. It is based on a collaboratively developed or 'shared' vision of desired future conditions, a vision integrating ecological, economic and social factors affecting a management unit defined primarily by ecological boundaries. It is adaptive in the sense that it assumes this vision will change over time in ways we cannot predict. It also recognizes that our current scientific knowledge, especially regarding ecosystem functions and responses, is (and will likely always be) limited. This, however, should be no excuse for inaction or indecision. We must act and make decisions today based on what we know today. As we learn more from our research we can and should adapt and alter our shared vision and decisions based on that new knowledge. The intention of this adaptive approach is to restore and maintain the health, sustainability, and biological diversity of ecosystems through a natural resource management approach that is fully integrated with and furthers social and economic goals.

This adaptive approach to environmental and ecological assessment, protection and management is not a new approach. Holling (1978) and his colleagues have been writing about and practicing this approach in various river basins throughout the world since the 1970s. Within the US government it has recently been rediscovered (or reinvented) under the title 'ecosystem' management. This reincarnation is somewhat broader in scope than envisioned two decades ago, and as such it has relevance with respect to long-term sustainability of water resource systems. This chapter will review some aspects of this approach with respect to water resource system sustainability.

The adaptive development and management approach emphasizes:

- Ensuring that all relevant and identifiable short and long-term ecological, economic and social consequences are considered.
- Improving coordination with and forming partnerships with all concerned stakeholders.
- Carrying out responsibilities more effectively and in a cost-effective manner.
- Using the best science currently available.
- Continually improving information and data management for decision-making.
- Monitoring the systems being managed and adjusting management decisions as new information becomes available.

This adaptive approach emerges from observations on how many public agencies (primarily) responsible for natural resource development and management have performed their missions. Past practices of those agencies have not often led to what might be called an integrated approach to resource management. Institutional constraints rather than individual limitations are often the cause. These constraints are now being recognized and attempts are being made to remove them.

The adaptive approach also emerges from the observation that people continue to want (and in some cases are willing to help pay for) the maintenance of a clean environment and biological diversity, while working toward economic prosperity. Laws and statutes increasingly promote multiple uses of natural resources and multiple rather than single objective mandates, and yet the resource management agencies typically remain responsible and have authority for only single, or at best only a limited set of, purposes and objectives.

Many practical advances in science and technology (e.g., modeling and data base availability and access) have enabled managers and the public to consider in their individual analyses many more variables over larger geographical areas and for longer time frames than was possible even a few years ago.

In many countries, however, there is continued frustration with protracted conflict and litigation and inaction, as well as decisions not favorable to the maintenance of desired natural ecosystems. This has led to the creation of a number of non-governmental organizations working to achieve more positive ways to resolve natural resource disputes and to promote sustainable uses of natural resources.

In short, this adaptive more comprehensive integrated 'ecosystem based' approach has been in response to the demand by many in the private sector, as well as in the public sector, for government agencies that work better.

The adaptive and more proactive approach to defining and implementing a more sustainable economy and achieving a more sustainable environment and ecosystem recognizes the connections between human and natural communities, and the ecological impacts of human actions. Everyone involved must be aware of the connections and use the latest scientific information that can help predict the impacts of selected actions. As new information and methodology become available, and as new goals emerge, planners and decision-makers must be prepared to adapt to them.

Governments are beginning to recognize that facilitating public/private partnerships is one way to increase public interest and involvement in, and knowledge of, decision-making. However, that in turn will require more intergovernmental cooperation, a more integrated and adaptive planning processes and a broader and longer-term perspective in natural (including water) resource decisions. These are all key elements of an adaptive ecosystem approach.

Where practiced, the adaptive ecosystem approach is viewed as having a number of benefits to all stakeholders, public and private alike. Its focus on social, economic and ecological sustainability can help draw stakeholders with diverse opinions and goals together. The approach helps national, state, regional and local governmental and non-governmental agencies and other parties work together to make and implement decisions affecting their ecosystems and environment, including those involving water resources. It combines conservation and economic goals of the private sector with restoration of ecosystems. The decision-making processes become more open to the public, with all interested parties encouraged to help establish project goals and to identify ways of achieving them. All of these are essential conditions for improving the sustainability of these systems.

An adaptive consensus-based resource management approach can also benefit the public by increasing incentives (as opposed to regulations) for moving toward ecological, environmental and economic goals. The avoidance, or at least reduction, of litigation and delays in the implementation of agency actions is a major benefit to everyone (except, perhaps to a few lawyers). With all stakeholders actively involved in the planning and implementation of a project, there tends to be less uncertainty regarding late changes in plans or policies. Collaboration on common problems also helps to ensure that there is balance and fairness in the consideration of the interests of small land owners and environmental interest groups. It also helps to ensure that all social, economic and environmental factors and concerns of all stakeholders are considered and judged with respect to the overall ecosystem.

To the extent that it works, the adaptive approach may also reduce costs associated with restoration of degraded habitats and their associated species populations. Duplication of efforts among various government and private organizations would also be reduced. The approach should also reduce those cyclical variations that adversely affect individuals and communities, events that have occurred in the past when planners have ignored environmental and economic issues.

While not suggesting it to be a panacea, the adaptive ecosystem approach is designed to promote a cooperative and participatory bottom-up approach involving all stakeholders such that it can help lead to a more sustainable use of natural resources, including those of water and related lands and ecosystems.

COMMUNICATION AMONG STAKEHOLDERS: FROM VISION TO ACTION

Stakeholder education and participation in decision-making is a central objective in attempts to implement an adaptive ecosystems approach. This includes involvement in the development and implementation of a shared ecosystem vision and strategy among all stakeholders. Public participation can include the design and implementation of processes for seeking public opinions, reviews of, and comments on, proposed governmental agency actions. Coordination of governmental agencies is essential when holding public meetings on the same subjects. Private organizations and individuals need to know which agency and person they should contact to get information that will make their inputs meaningful. Communication, often relegated as a part-time job of individuals in government agencies, is in reality a full-time job. Individuals responsible for the communications can often benefit from specialized training. For one, they need to learn how to effectively communicate technical information to the interested and informed public, as well as to managers. This can only be achieved if such information is presented in ways that those not specifically trained in the applicable discipline(s) can understand. It can be done, as illustrated by a few USA examples below.

Example 1: Cordova, Alaska (USA)

As a result of litigation following the Exxon Valdez oil spill (Alaska, USA), a court-ordered consent decree (settlement) established a state and federal interagency trustee council to respond to the spill. The council's plan outlined how to achieve restoration of the oil spill area and how to spend the settlement funds. The vision – restore the resource affected by the spill – is required by law. The council formalized the broad objective into a strategic plan for implementing the vision. To illustrate what could be done, the Prince William Sound Science Center was established by the people of Cordova, Alaska. The Center developed a cooperative education program with state and local agencies and with the local school district. This program shares research and geographical information system (GIS) information with the local community. The Center gives people access, in one central location, to a wide array of information on the ecosystem, including its current state. In addition, the information is being put on-line for public access via personal computers.

Example 2: Anacostia, Maryland (USA)

Anacostia, on the east coast of Maryland (USA), has a Watershed Restoration Committee. The Committee included representatives from the District of Columbia, two counties and one state. The US Army Corps of Engineers represented the federal agencies. Facilitated through the Metropolitan Washington Council of Governments, this committee established a six-point action plan for restoring the Anacostia River basin. The plan identified agencies involved in the restoration effort, described proposed and completed projects and outlined problems, strategies and challenges associated with achieving the goals. The plan defines a vision by local and state governments. Federal agencies were perceived as facilitators and implementers of local goals. In this case, the consensus vision that was developed provided a general guide for diverse entities to move in a common direction.

Example 3: Connecticut wetlands (USA)

The restoration of some northeast USA coastal wetlands and estuaries was also guided by a vision of opportunity developed by all stakeholders. A primary railroad between New York and Boston bisected Connecticut coastal wetlands that at the time of construction were considered undesirable. The railway embankment also restricted the tidal flushing of these wetlands, and over time the wetlands on the inland side lost much of their value to finfish and shellfish. The rebuilding of the railway corridor to help meet future demands provided an opportunity simultaneously to reintroduce

tidal flow to the degraded wetlands, leading to the vision of healthy and productive coastal wetland habitats.

Example 4: Cape cod, Massachusetts (USA)

Similar restorative efforts are underway in Waquoit Bay on Cape Cod, Massachusetts, USA. Land use patterns there have changed over time, from 2 percent residential in 1950 to 20 percent in 1990, with a 15-fold increase in human population leading to increasing stress on the watershed. Collaborative efforts among federal, state, regional and local governments, as well as environmental groups and the general public, has resulted in water quality monitoring by volunteers and scientists, the acquisition of a No-Discharge Order for the bay, and serious discussion concerning the creation of a 2,500-acre wildlife refuge that would prevent development and protect it for use by wildlife.

The above examples are only a small sample of what could be presented to show that people with diverse interests can get together and work toward a common vision if they are able to conceive the vision. Communication is essential. Technical aspects need not be roadblocks, in fact they can actually facilitate and focus discussions.

DEVELOPING A SHARED VISION

Achieving a 'shared' vision of any particular water resource system, its watershed, and its ecosystem is an important step toward reaching decisions that will promote sustainability. It can be facilitated by involving all stakeholders in the development of the predictive simulation model needed to generate the time series of future values of the various performance indicators, including the ecological variables, associated with any particular system development or management policy. The desired outcome of such an exercise is a highly interactive computer model that is developed jointly by stakeholders in a region and that contains a collective view of the system and how it is operated. It is a model that:

- all agree represents the system,
- all agree can be used to derive a development or management policy,
- all agree adequately predicts the impacts of concern,
- is trusted by those engaged in the planning exercise.

To achieve this, all stakeholders who have interests and influence on decision-making must be involved in this shared vision exercise. The exercise must have a well-defined purpose. There must be a willingness to cooperate among all stakeholders – i.e., a desire to 'get to yes'. Finally, it helps to have a facilitator who has experience in working with diverse interest groups, who has prepared material that will help everyone learn how and gain confidence in the model building technology itself. Having all the necessary data available during the shared vision model development exercise is also helpful, but part of such exercises can be focused on identifying just what data are important, or sensitive to, the decisions being considered.

Shared vision models, once developed, can be used for joint fact finding, problem identification, assumption and constraint clarification, group brainstorming (trial and error), tradeoff identification and assessment, strategy refinement and communication, and the maintenance of changing (adapting) development plans and policies. Shared vision modeling has been facilitated by the development and use of generic object-oriented simulation 'shells'. Using the graphics interface of these computer programs or shells, one can 'draw in' their node–link representation of the water resource system and then interactively enter the necessary data associated with each 'node' and 'link'. Once the data have been entered, the defined system can be simulated. The resulting output can be displayed over maps and graphically as well as numerically. Based on the results, the users of such shells can easily modify their data and re-simulate. This iterative interactive trial-and-error experimental process can help all stakeholders learn more about their system and how it might best be developed and managed.

Shared vision planning for water and ecosystem management also involves stakeholder participation. Part of this process may involve conflict resolution in order to create a common understanding and vision among all stakeholders. The shared vision exercise involving computer simulation modeling should engage participants in a disciplined approach to defining planning objectives, developing teams, creating measures of system performance, defining the status quo, evaluating alternatives, and finally in the creation and implementation of the plan.

As much as the adaptive ecosystem approach requires a holistic view of the ecological and socio-economic aspects of an entire landscape, it also requires an integrated institutional and fully participatory planning and decision-making process. Achieving a common vision of the desired ecosystem health and economic indicators and the indicator values is much more difficult than individual economic performance measures. But the shared vision modeling exercises can help in communication among the diverse interest groups. Experiences in carrying out shared vision modeling exercises point to their potential value in achieving a common understanding and shared sense of purpose among individuals who may not all have the same interests and concerns with respect to the development or management of a water resource system, a watershed, or river basin (Holling, 1978).

RESOURCE ALLOCATION AND MANAGEMENT

Decisions about the allocation and management of resources, money, people, time and even equipment provide a useful measure of agency or organization identity. Managers tend to allocate their resources to those areas closest to the central mission of their organization. To be effective, however, interagency coordination in an ecosystem context also requires some degree of budget coordination among agencies. Managers should not be asking how to fund ecosystem management in addition to traditional activities, but rather how to implement their mandated activities using an adaptive ecosystem approach. This may involve determining which traditional activities can be incorporated into ecosystem efforts and which can be ignored.

As an example, the Federal Bureau of Land Management office in the state of Idaho (USA) recognized some time ago that its jurisdictional boundaries in Idaho were not correlated with ecological systems. The Bureau thus reorganized itself into a team structure focusing on ecosystems. This significantly reduced the need for personnel in the State headquarters and put more personnel into the field offices. As a result, the Bureau estimates a 30 percent increase in efficiency and productivity for its Idaho operations.

KNOWLEDGE BASES AND SCIENCE

Sciencie is a particularly crucial component of the adaptive ecosystem approach. Science is needed to describe the state of the ecosystem, to assess its vulnerability to stress, to identify ecosystem processes needed to achieve various objectives, to establish restoration techniques and to monitor effectively any ecosystem changes. Too often, however, information about ecosystems of interest, and levels of understanding about ecosystem functions, are inadequate for ecosystem analyses and predictions.

The adaptive ecosystem approach also requires scientific information on how people interact with natural resources, how their plans and aspirations relate to natural systems and a host of other economic and sociological questions.

Economists and sociologists are necessary for the understanding and integration of ecological, social and economic concerns. The approach requires multi- and interdisciplinary inputs as well as research.

Unfortunately, each scientific discipline tends to identify and communicate almost solely with members within that discipline. Individuals look to their disciplinary professional associations for the exchange of ideas and outlets for publication. Public agencies tend to hire those whose disciplines most closely match agency missions. And because the agency engineers and scientists are career employees, it is often difficult for the agencies and organizations to move quickly into new disciplinary areas as needs and goals change. (The US Army Corps of Engineers, for example, a large public resource management agency, has recruited from many disciplines. Its mission is very multi-disciplinary. However, the Corps recruitment of other than engineers has generated criticism from some engineers who see a diminishing role for their discipline within the Corps 'of Engineers'.)

There are a number of challenges facing those who are attempting to generate and communicate a base of scientific knowledge that they believe will help enable the implementation of an effective ecosystem approach that involves many disciplines and the public:

- Engineers and scientists tend to publish their results in professional journals typically read only by other engineers and scientists.
- Reports of scientific studies are usually narrowly focused, making them difficult for managers to use. Managers must focus on larger issues than those of interest to most scientists.
- Scientists strive for precision, while managers must often make decisions based on uncertain or scarce information available.
- Scientists often focus on the lack of information as a justification for new studies while managers must focus on whatever information is available as a basis for their decisions.
- Scientists and engineers work to reduce uncertainty where they can; managers and decision-makers tend to exploit uncertainty, especially if it leads to decisions they want to make anyway.

All these factors are worthy of attention by those who are going to join and lead in the effort to change the normal way of organizing themselves and doing things in ways that may be viewed by their descendants as having led to more sustainable water resource system designs and management policies.

INFORMATION AND DATA MANAGEMENT

Access to accurate, up-to-date and comprehensive information is essential for effective decision-making at an ecosytem or regional level. Common access to the same information provides a common basis upon which all participating organizations and stakeholders interests can begin. Normally no single public or private agency or organization has the resources or the mandate to collect or maintain all relevant data on any ecosystem. Thus, an information-rich understanding that spans the entire ecosystem almost always requires the combined efforts of many agencies and institutions. Access to that information should be made uniform among all the stakeholders.

Without a coherent and complete picture of the resources affected by their decisions, managers may unwittingly take actions that degrade the ecosystem. If managers find themselves having independently to create the information they need so as to avoid such actions, duplication of effort is likely. Cooperation, coordination and communication are essential for any ecosystem approach to succeed. This will not happen without it being an important goal within each stakeholder group and without considerable effort to achieve that goal.

Modern computer technology permits improved data management and sharing among all interested parties and organizations. Information should be focused on key indicators of ecosystem functions. Information should be standardized with respect to terminology, definitions, procedures and geographical references. It should also be subjected to quality controls. This becomes especially important when multiple parties are accessing and using these data together with models, possibly 'downloaded' from electronic networks using, for example, the World-Wide-Web (WWW) and its associated software. All participating agencies and organizations should have access to at least some of these facilities.

HYDRO-ECOLOGICAL MODELING

There is an increasing number of attempts to model ecosystems so that any impacts on such ecosystems by alternative water management policies can be predicted. Hydro-ecological (or eco-hydrological, depending on the relative emphasis) modeling attempts to predict the consequences of a change in water management on associated ecosystems. Currently, many predictions are carried out but for only limited geographic regions. Furthermore, the focus is primarily on the vegetation. This is done because of the important role of vegetation in the ecosystem, because vegetation is a directly influenced part of the ecosystem and because vegetation is the simplest component to predict.

These eco-hydrological or hydro-ecological modeling activities are in the beginning stages when compared to, for example, water quality modeling. If successful, they will be of considerable value to land use and water managers, who need to know what the impact of their land use and water allocation and wastewater control decisions will have on a water resource system's ecology.

Most hydro-ecological models that have been developed thus far tend to rely on decision rules for, and regressions of, vegetation/hydrological relationships for specified soil types and conditions. These rules are usually based on expert judgment and statistical analyses of field data. Modeling wetland responses, for example, involves combined surface and groundwater quantity and quality modeling. This is not a trivial exercise, even if it is limited to the hydrology alone. Wetland ecology may also be influenced by such additional factors as:

- the surrounding ecosystems and land uses,
- the type of soil and its sediments and chemistry,
- water levels and their fluctuations,
- water chemistry including its characteristics of salinity, nutrients, acidity or alkalinity,
- oxygen and other possible constituents (some toxic) and their fluctuations,
- the groundwater flow, with influences on carbon-oxides,
- direct human influences, such as application of pesticides and other water and land management decisions.

To construct a model that links water to vegetation or aquatic animal specie groups, or vice versa, some cause–effect levels need to be defined. Each level is associated with certain types of data and an apropriate degree of detail necessary for subsequent levels. Figure 7.1 illustrates the interdependence among major modeling components of human choice and concern.

Models must link land and water management modifications to changes in regional hydrology. For this one or more hydrologic models are needed. Ground water is often of concern, and if so, a geohydrological model is needed. The change in regional hydrology must then be translated to more local site-specific ecological effects. This step could be called eco-hydrological modeling to the extent that it supplies data for predicting the impacts of changes in the local hydrologic regime (such as quantity and quality and depth of water flow in and above the soil) and its impact, in turn, on the health and vitality of important vegetation and animal

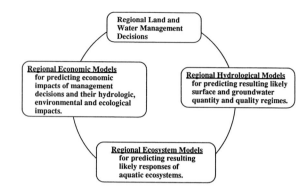

Figure 7.1 Major interdependent or linked modeling components for predicting economic, hydrologic and ecological impacts of alternative land and water management policies and uses.

types. Such predictions are clearly dependent on other factors distinguishing the particular site, such as existing or changed land uses not directly predicted or modeled, atmospheric deposition and other characteristics.

Finally, it is of interest to most of us to know what, if any, economic consequences may result from possible alternative land and water management decisions together with their hydrological and ecological impacts. Will a proposed land use or water management decision, and its resulting hydrologic, environmental and ecological impacts, create increased economic and social demands and activities in the region, and if so of what kind and to what extent? Will the development of a community shopping center on a portion of an existing flood plain, for example, adversely impct the ecological and hydrological integrity of surrounding wetlands? will increased commercial use of this land increase pressure for further wetland drainage and development, thereby possibly increasing the potential for increased pollution and risks of flood damage? Questions such as these are often of interest to those in positions of authority who must approve or disapprove requests for land and water use changes. Regional economic models that are sensitive to land use activities and the hydrology and ecology of the region are needed to make such predictions.

FLEXIBILITY FOR ADAPTIVE MANAGEMENT

Decisions regarding ecosystem management are always made with some degree of uncertainty or risk with respect to the resulting impacts. It is unlikely that we will ever have enough understanding and information to eliminate this uncertainty. Hence, there should be a monitoring program established and funded, and a periodic review based on the data collected. If upon review of these data, or based on other information, a shift in management policy is warranted, it should be taken. This new policy should be monitored, just as before. It too should be reviewed periodically to determine if further changes are desirable. This is called *adaptive management*.

Adaptive management is a continual cycle of planning, acting, monitoring, evaluating and back to planning. Adaptive management requires ongoing testing of management choices. Such testing must be based on a systematic program of design, research, monitoring and evaluation. A successful adaptive management program requires an organizational commitment for an extended period of time as well as an ability and commitment to deal with the setbacks and frustrations that are unavoidable consequences of experimentation.

SOME THINGS TO REMEMBER

The important major steps in the implementation of a comprehensive, multi-sector, adaptive ecosystem approach include:

- Develop a shared vision of the desired ecosystem condition that includes social and economic conditions and identify ways in which all parties can contribute to achieving stated goals.
- Develop coordinated approaches among all concerned and interested agencies to accomplish ecosystem objectives, collaborating with all stakeholders in recognition of mutual concerns.
- Use ecological approaches that restore or maintain biological diversity, health and sustainability of the ecosystem.
- Support actions that incorporate sustained economic, social-cultural and community goals.
- Respect and ensure private property rights and work cooperatively with private landowners to accomplish shared goals.
- Recognize that ecosystems and institutions are complex, dynamic, changing and characteristically heterogeneous over space and time.
- Use an adaptive approach to management to achieve both desired goals and new understanding of ecosystems.
- Integrate the best science available into the decision-making process, while continuing scientific research to improve the knowledge base.

- Establish baseline conditions for ecosystem functioning and sustainability against which change can be measured. Monitor and evaluate actions to determine if goals and objectives are being achieved.

As different organizations gain experience in working together towards implementing an adaptive ecosystem approach they will discover what works, and what does not work, for them. For those involved for the first time, an adaptive approach is offered in the following paragraphs. The order in which the guidelines within this framework are adopted may differ for different groups, and not all of these guidelines may be applicable for every ecosystem initiative.

- *Define the area of concern or of interest.* The area boundaries may be influenced by a number of issues, including economic, social, cultural, ecological and political. If the issue is the management of water resources, it should be viewed in the broader ecosystem context. How the issue is framed will determine in part which stakeholders will become involved. The initial issue of interest, e.g., water resources management, may expand to include many related issues, which in turn will interest other stakeholders. The boundary size should be large enough to be meaningful, and small enough to be focused and allow effective actions by the participants.
- *Involve stakeholders.* Involving all stakeholders is an important component of the ecosystem approach. Governmental agencies and other private organizations should provide sufficient financial support where needed to ensure that all stakeholders can remain involved.
- *Develop a shared vision of the ecosystem's desired future condition.* A vision statement is a clear conceptual picture of the desired future state towards which efforts are directed. It defines the principal benefits to all stakeholders. It should be consistent with the overarching goal of maintaining the good health, sustainability and biological diversity of the ecosystem while supporting communities and their economies. Vision statements should be both broad and precise.
- *Characterize the present ecosystem including its cultural, economic, environmental and social conditions and trends.* This provides a baseline for measuring what restoration efforts are needed and any progress made. The ecosystem can be characterized by such variables as composition, structure, function and natural range of variability for key ecosystem characteristics and ecological stresses such as toxic pollution. The social environment can be characterized by such variables as location, distribution

and size of communities, human uses of ecosystem resources and political and economic issues that have a bearing on resource use. The economic environment can be characterized by such variables as local employment patterns, work force availability and skills and location and distribution of important economic centers.

- *Establish ecosystem goals and objectives.* Goals and objectives can be expressed as desired results. These results should be achievable and measurable so that one can measure progress and eventual success. Objectives should be quantifiable, verifiable and flexible. Mechanisms for resolving inevitable conflicts over multiple objectives need to be established from the beginning.

- *Develop and implement an action plan for achieving the specified ecological and environmemtal goals and objectives.* An action plan specifies detailed steps to achieve goals and objectives. These actions may include engineering, administrative, financial, scientific and numerous other types of activities. The plan should also provide for coordination of the various active interests, and for obtaining public involvement and comments from groups and individuals that are not participating as stakeholders.

- *Monitor conditions and evaluate results.* Monitoring can provide information to determine if and the extent to which the specific objectives and goals are being achieved (and if the underlying assumptions are correct). Monitoring is particularly important when new or unproven methods or technologies are being applied, when significant levels of risk and uncertainty prevail, and/or when it is necessary to determine whether management or restoration measures are working as planned. Monitoring should detect changes in ecological, social, cultural and economic systems; provide a basis for natural resource and other policy decisions; track status and trends; compile information systematically; link overall information strategies for consistent implementation; and ensure prompt analysis and application of data in the adaptive management process.

- *Adapt management according to new information.* Adaptive management is a process of adjusting management actions and directions, as appropriate, in light of new information on the condition of the ecosystem and progress toward meeting the goals and objectives desired. Management decisions are thus viewed as experiments, subject to modification – but with the goals clearly in mind. Adaptive management recognizes the limitations of current knowledge and experience, that we learn by experimenting, and helps us move toward goals in the face of this uncertainty. It accepts that there is a continual need to review and revise environmental and other restoration and management approaches because of the dyanmic as well as uncertain nature of ecosystems.

8 Institutional and social aspects of sustainability

Actions taken to increase the sustainability of our economy and society are social actions. For, while these actions may indeed be based on knowledge from, and methods of, engineering and technology, economics, anthropology, planning and law, they remain human activities, and hence they are social. Water is an essential part of any economy and society. The sustainability of a water resource system is a necessary condition for a sustainable economy and society. But it is not just the liquid part of the water resource systems that are trying to be made more sustainable. It includes the physical–technical–social sub-systems and the capacities and capabilities of those individuals who design them, who manage them and who use them. It is these social components of systems that are often the most challenging because they involve changing the way individuals think and behave. Fortunately, however, society is adaptive – its institutions and individuals change over time in ways that can allow society to work better.

How can society change to one that thinks more 'sustainably' – to one that will evolve toward a more sustainable state? Society, through its organizations and demands, makes possible the development of science and technology that are important determinants in its attempts to increase the quality of our lives, our economies and our environment. Science and technology together with our institutions can permit us to have more while consuming less. Society and its institutions shape the physical systems – e.g., the water resource systems – that are built or managed to serve us, and those systems in turn shape and have impacts on society. The complexity of the interactions between society and its physical–technical systems is a major factor frustrating many who are working and contributing their talents toward the solution of social problems – problems that seem to limit how far and how fast society as a whole can move toward increased sustainability. Without the institutions that can design, develop and manage our water resource systems, their existence or function would be severely constrained. It is the institutional constraints, not the technological ones, that most often prevent a more effective and efficient use of these resources.

SOCIAL RESPONSIBILITIES

Do the water resource systems we develop and manage serve to enhance the long-range sustainability of our societies? What are the social responsibilities of water resource planners and managers, and indeed of engineers and scientists in general? These are two questions that should be addressed whenever we as managers or planners make decisions with respect to our professional responsibilities. For example:

- Do we consider and acknowledge the importance of individuality, privacy, diversity and aesthetics? If so, how do we make tradeoffs among these and other criteria or objectives.
- Do we avoid dangerous or uncontrolled side-effects and by-products (and articulate and communicate our concerns to the public) when faced with a multitude of pressures to do what we consider is inappropriate with respect to the best interests of current or future populations and societies?
- Do we make provisions for consequences should the systems we design and manage fail? If so, are they adequate given the risks of failure, and how can we determine this?

- Does our ability to apply a multitude of quick technological fixes cloud our ability to work toward fundamental change, when needed, in the way we design and manage our water resource systems? Where are fundamental changes needed?
- Do we as water resources planners and managers become involved not only in defining what we do, but also why we do it?
- Do we as professionals continually strive to serve society as a whole and not just our own special interests and purposes? How can we be sure of this?

If we as professionals are to have a more decisive and enlightened social role, we must be willing to:

- work more closely with leaders of society to develop a sense of what is right with respect to the planning and management of water resources,
- engage more actively in the political dialogue and in the definition of what actions are needed to increase the sustainability of our current and future water resource systems,
- reshape and improve the liberal as well as technical education of both professionals and others in the use of water resources to better serve society, and
- encourage the involvement of professionals from all disciplines in the planning and management of our natural resources, and become engaged in all aspects of society to broaden our views of what is or is not more sustainable.

Most of us as individuals, whether professional or not, work within social and institutional organizations. We function under rules or procedures established to achieve order and predictability. These rules and procedures may be formal and codified, as they are in constitutions, statutes and regulations. Alternatively, they may be informal and implicit, as are many of the rules that govern relationships within a family, or within an organizations or firm. These rules obviously are always subject to intervention and change.

Institutions, i.e., these sets of rules, are both instrumental and nested. Individuals generally operate within several institutions simultaneously. The consideration of sustainability criteria along with the more common economic, environmental and social criteria may point to the need to change how we develop and use our resources. Any process involving change will require that we change our institutions – the rules under which we function. This is certainly one of the most difficult, yet most important, of the challenges we all face in attempting to move to a more sustainable economy and society. These are serious challenges. Social scientists

studying and analyzing institutional mechanisms that enable us to change are involved in what is really the 'hard' science.

INSTITUTIONS AND CHANGE

Institutions can be divided into various levels. One division distinguishes three levels: the operational, the organizational and the policy levels. The rules at the operational level are those that govern the behavior of individuals. Rules at the organizational level determine the rules at the operational level. The rules at the policy level are those that determine the rules at the organizational level.

An example of this hiearchical arrangement, as shown in figure 8.1, might be a permit system for water use. At the operational level water users could be required to have permits specifying the conditions under which they can use water. The implementation of such a permit system, as well as the contents of particular permits, could be the function of a water agency at the organizational level. The rules under which such a water agency operates, i.e., how they issue permits, might be determined by a governing body at the policy level.

The notion of institutional levels makes it possible to address questions of power and collective action. Viewing the higher levels of institutions as sites for rule-making that affects or applies to lower levels is one way to work through notions of differentiated institutional positions, and therefore differentiated power relationships. Those with positions and authorities at higher levels of institutions may not set the specific rules for operational level actors, but they set the rules for making rules. The distinctions drawn about the different actors operating at different institutional levels

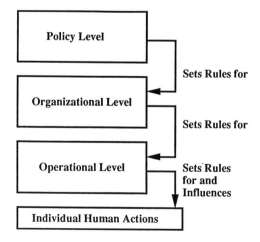

Figure 8.1 Institutional or organizational levels and functions.

adds needed specificity. It allows an examination of the critical issues concerning which actors, which skills and which temporal and spatial structures one chooses to investigate and/or modify. Sustainability criteria could be established for each level. These criteria would likely be relatively broadly defined at the higher (more strategic) institutional levels and more specifically defined at the lower (more tactical) institutional levels.

The criteria that are applied for understanding institutional levels (in other words, how decisions are actually made) can be unforeseen and changeable. Why? Because as complicated as they may sometimes seem, the situations are seldom as simple and straightforward as they may seem, especially to an 'expert' who is not aware of local situations. For example, in a study done for the United Nations Economic Commission for Latin America and the Caribbean (ECLAC, 1989) on several countries of the region it was noted that the criteria will escape the unconscientious researcher who limits himself to evaluating the functioning of an organization only through its organizational structure, work plans, official statements, or progress reports.

> The reality is that behind each organizational structure there are other organizational structures which are invisible to the unconscientious observer, made up of commissions, work groups and groups of friends, unspoken agreements and rules of play, special interests in some zones of the country, and other factors which, although they may be logical to each actor or group of actors, can lead to situations exactly the reverse of the explicit government positions. In light of this, the detection and constitution of a list of the criteria governing water systems management requries a meticulous research effort. This is important and necessary, because these criteria can explain much more clearly this behaviour of each actor involved in the management process.

Thus, to understand fully the boundaries of relevant institutions, the planners must be very careful to be sure they understand the 'real world'. But they will also want to know, and are usually interested in how institutions function under stress, or under pressures for change from individuals within and outside the institution. Any time there is dissatisfaction with the status quo there is a gap between what exists, or how individuals and their institutions behave, and how at least some wish they would behave. This dissatisfaction gap, once identified, may stimulate action. Such action may take the form of changing the rules at any of the three institutional levels. Individuals will adopt strategies (often through the formation of coalitions) aimed at closing the gap. It is, of course, possible that some dissatisfied individuals may seek remedies to problematic situations by going outside the rules altogether.

If complaining individuals, or groups of individuals, work within the existing institutions, one element of such strategies is likely to include a request to policy-makers to change the rules of the lower-level institutions. The complaining individuals might specify the rule changes that they want adopted (if these can be articulated) or they may emphasize the gaps that they want closed. In either case, they place a demand upon higher-placed individuals to act for them. These higher-placed individuals, in responding to such demands, may or may not pursue a process of institutional (rule) change.

In any case, the process of institutional analysis begins at this point, when policy-makers (i.e., those in positions to change rules), in response to policy demands, call upon analysts to advise on the options available and their respective advantages and disadvantages. They will also want to know who and how many will care, and how much they will care. The task and challenge of a policy analyst is to prescribe or recommend specific rule changes (institutional innovations) that, if adopted and implemented, are likely to lead to closure of the dissatisfaction gaps which gave rise to the policy demands.

These rule changes are the outputs of the institutional change process. Institutional change may also have other consequences, some direct and some indirect, some intended and others unintended. All are what many analysts refer to as 'outcomes'. Individuals then weight these outcomes, and decide whether they are satisfactory or whether they leave the initial problem unresolved, or even create new ones. In the latter two cases, which are quite common, the process of institutional change becomes an iterative one.

It is the individuals' abilities to monitor these outputs and outcomes and to perceive and evaluate the nature of these effects, that determines their participation in subsequent rounds of institutional transformation. If individuals are satisfied, they may then work to preserve the new status quo. If they become aware of inadequately addressed gaps, new gaps or undesirable outcomes, they may participate in a new round of rule changes.

Institutional change occurs quite often in water management. Clearly any move toward the planning, management and design of more sustainable water resource systems will also involve institutional change. Such changes are already occurring within some professional engineering societies throughout the world. They are also occurring at various levels of national and state governmental agencies, as they adapt to a more integrated holistic and sustainable approach to water and land management.

Three examples of institutional change, and the resistance to change, follow.

WATER MANAGEMENT IN THE RIO GRANDE (USA AND MEXICO)

Water management institutions are often under pressure for change as a result of changing demands for water and/or some of the purposes water resource systems can serve. Who gets how much, when, in what condition or just how much flood protection is provided to whom or where, or just what the water quality conditions should be when and where are all decisions institutions must make. Consider, for example, the Rio Grande River in southwestern USA (known as the Rio Bravo del Norte in Mexico). This river (as partially shown in Figure 8.2) is approximately 3,000 km long and flows from southern Colorado through the states of New Mexico and then becomes an international river between Texas (US) and Mexico. It drain an area of over 900,000 km^2, over half of which is in Mexico. It is the lifeblood of this semi-arid region, supplying water to over 3.5 million people as well as providing hydropower, irrigation, industrial and recreational benefits. Over recent years there has been a definite shift in demand from agricultural to urban, environmental and recreational uses.

Research reported by Waterstone (1994) addresses the potential of institutions involved in the management of the Rio Grande's waters in the USA to adjust to these changes, and to possible changes in demands as well as supplies due to climate change. What happens if the stresses go beyond the ability of the existing institutions to manage? How will the rules be changed to accommodate the possible gaps between what now exists and what is desired, especially if what is desired by various individual water users or interest groups are, when all the desires are considered together, impossible to satisfy?

At present the physical system of the portion of the Rio Grande in the USA comprises a series of eight major linked reservoirs operated principally by the US Army Corps of Engineers and the US Bureau of Reclamation. The reservoirs are managed to fulfill a variety of purposes including flood control, municipal and industrial water supply, irrigation, hydropower production, sediment control, environmental enhancement and recreation. As with many multio-purpose reservoirs, these are managed to strike an appropriate balance among varying, and sometimes conflicting, uses.

In this semi-arid region, a continual concern is the availability of water in storage to meet future irrigation demands.

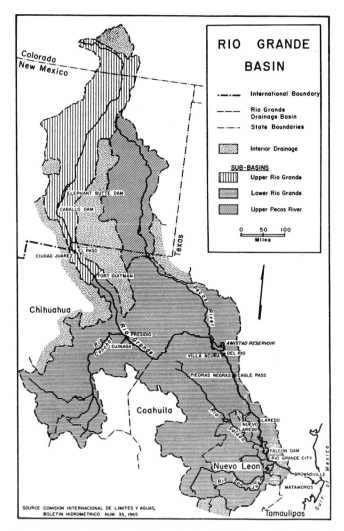

Figure 8.2 The Rio Grande River Basin in Mexico and the United States of America.

However, there is also a need in these water supply reservoirs to preserve space for accommodating flood waters and preventing downstream flood losses. To accomplish these various purposes, the reservoirs are managed, on a day-to-day basis, under a series of complex rules. In terms of the framework described above, these rules constitute the operational level of the water management institution for the Rio Grande basin. These rules are nested within, and are compliant with, a set of organizational-level rules and a set of policy-level rules.

This characterization of the water management system of the Rio Grande enables one to understand how operation-level rules are created and how they might be changed if they become deficient. At the policy level, there are water delivery requirements mandated under the Mexican Water Treaties of

1906 and 1944 that allocate the water supply of the river between the USA and Mexico. Another is the set of water deliveries required by the Rio Grande Compact of 1938 that allocates water among the three USA basin states: Colorado, New Mexico and Texas. In addition, at the policy level are rules specified by the Reclamation Act of 1902, the Appropriation doctrine of 'first in time, first in right', the Federal Reserved Water Rights for Indian tribes and other purposes, various environmental and water quality laws, and the water laws of the three basin states. These policy-level rules specify the implementing-organization level actors, indicate the scope of their activities and impose responsibilities and confer authorities upon them.

Organizational-level rules specify precisely how, and under what conditions, the reservoirs and other facilities should be operated. The actors at this level (principally the US Army Corps of Engineers, the US Bureau of Reclamation, the basin states' water agencies and the major water users), working within the rules from the policy level, develop the strategies and mechanisms. These are the operational-level rules for day-to-day operation of the structural and non-structural features of the Rio Grande system.

Operational-level rules are specified daily, based on a diverse set of formal and informal rules which emerge from the two previous levels. These include the standard operating procedures for the reservoirs under various conditions, the system of water rights for users, water quality permits and many others. While rules from the higher levels specify who may use water and in what ways, the actions that reservoir operators actually take are specified by rules at the operational level.

WATER MANAGEMENT ON THE YANGTZE (PR OF CHINA)

China's Yangtze River (Figure 8.3) is the third longest in the world. Only the Nile and Amazon are longer. It is the source of water for much of the rice and fish production in China, and it is also the cause of major damages and deaths. China's engineers and policital leaders have been developing plans to harness the river, reduce its floods, provide for navigation, and produce hydroelectric power throughout this past century. Now, at the end of this century, they are initiating a key component of those plans, the building of the Three Gorges Dam. If built as currently planned, it will be the world's largest dam – over 180 m high, spanning some 2 km, and creating a lake some 600 km long and 180 m deep. Its hydropower turbines would be capable of producing over 18 thou-

Figure 8.3 The Yangtze River in the People's Republic of China.

sand megawatts of power, or some 80 billion kilowatt hours of electricity each year. Deep draft ocean-going freighters would be able to travel some 2400 km inland, and the downstream damage and loss of life caused by floods originating upstream of the dam would be eliminated. These are the plans and expectations of those who advocate the implementation of this project (Spence, 1997).

The estimated US $30-billion cost of this project will top the expected full-development cost of the Great Man-Made River Project in Libya, becoming the world's most expensive civil works project. It will also force the relocation of about 2 million individuals who now live where the lake will form. Archeological sites will be inundated and aquatic species will be lost. It will permanently alter what is now considered some of the most spectacular scenery in China. Heavy siltation so characteristic of the Yangtze and water quality problems due to the discharge of point and non-point sources of untreated pollutants may be significant. In addition there is the worry about the consequences of a dam failure, should it occur. The project's protential adverse cultural, environmental and social impacts have substantially reduced the supply of foreign financial aid and technical assistance typically available and provided by international development banks and individual countries.

The history of the planning of this project reflects a classic conflict over economic and political power. Those favoring the project have included the Yangai Valley Planning Office whose leadership has close ties to the government of Beijing. By the late 1950s the planning staff numbered about 12,000. One of their goals was to permanently remove the threat of floods like that which occurred earlier that decade, in 1954. Different plans were advocated by others in the government. One, proposed by the ministry of electric power, favored a group of smaller dams on the tributaires of the Yangtze, at least as a first step in developing increased energy and reducing expected flood damages.

A vote taken in the People's Congress in 1992 showed that the majority supported the Premier who is a strong advocate of this project. Construction on the massive project began in 1993.

In a country where conflicting political issues are not always accompanied by a full public debate, on this issue there is a surprising number of critics who have been able to express themselves. The debate continues in spite of the project's official approval and implementation. Both Chinese and foreign observers of this undertaking have pointed to the current institutional bias towards huge engineering projects as a way of solving long-standing water resource management problems. They have also observed that those politicians who support the project also support authoritarianism, central economic control and the ultimate authority of one person. Those who disagree are typically imprisoned. Those currently opposed the dam are mostly intellectuals who are concerned about adverse financial, environmental and ecological impacts as well as human rights and the preservation of historical and cultural assets. They are worried about the potential for the government to make another mistake similar to the Great Leap Forward and the Cultural Revolution. They are worried about dam failure – given China's dam failure rate that is some five times the world's average. They point to the failure of the Henan Province dam in 1975, caused by poor construction and a record typhoon, that killed over 250,000 people and caused disease and famine for some 11 million more.

China, with about a quarter of the world's population, has enormous problems, not just those related to water. Will the cultural, economical, environmental and social benefits of this project exceed the costs? These issues have to be debated and all those who are impacted, i.e., the entire country, should be involved in this debate. No one single institution or individual has the wisdom to know and dictate what is best for improving the quality of life for those now living in China and for those who will be living in China in the future. While a single institution or individual make make the final decisions regarding the continued development of this major project, it should make this decision with all the information

it can get. This can only come from a full and open debate of all the issues and tradeoffs that are involved. It could be that this project will prove to be a wise move toward increased sustainability, but unless some of the potential problems already identified are adequately addressed and debated, it could also turn out to be a step towards short and long-term unsustainability.

THE GABCIKOVO DAM CONTROVERSY (HUNGARY AND SLOVAKIA)

The controversy between Hungary and Slovakia continues over the diversion of the Danube River in Central Europe by Slovakia to feed the recently constructed Gabcikovo hydropower dam. This dam together with a companion dam were conceived in the early 1950s by both Hungary and what was then Czechoslovakia. The agreement, signed in 1977, ignored environmental and ecological impacts, as was the habit of Eastern European governments at that time. Then in the 1980s pressure from environmental interest groups resulted in Hungary's canceling their participation in the two-dam project, leaving only one, Gabcikovo, that was built by the Slovaks. Currently some 48 km of the Danube river, which originally formed a border between Slovakia and Hungary, is now located in Slovakia. Uncertainty over potential future ecological and environmental impacts, the border issue, and long-standing ethnic hostilities all reflect major changes in the geopolitics of this area over the past decade. The public's backlash in Hungary against those responsible for the planning of this project, a symbol of many such environmentally-blind water resource development projects implemented under the previous government, and especially against the engineering community, is still evident.

More importantly, the conflict underscores the current lack of effective institutions within Hungary and Slovakia that can manage this international as well as intranational conflict. As a result, both countries have submitted their conflict to the International Court. The adjudication of this case will in all likelihood take years. Meanwhile both countries have expressed the desire to join the European Union. This has increased their willingness to resolve the conflict. As members of the European Union they will be subject to non-binding Recommendations and binding Directives. The latter commits all member countries to enact legislation or institutions that will ensure the directives are carried out. Many of the directives issued thus far involve water resources quantity and quality management issues.

CONCLUSION

One of the most difficult parts of a water resources system to change are its institutions that plan, design and manage the water itself. But to the extent that sustainability criteria identify needed changes in the way we develop and manage our water resource systems in the future, this will involve not only physical and technical changes, but institutional as well. The events taking place in the Rio Grande, the Yangtze, and the Danube river basins are just three examples illustrating the strong links between the technical and institutional components of water resource systems. The more we can understand these links and just how institutions function within our own regions of interest, the better we will be able to work with those institutions to achieve more sustainable water resource systems. Understanding how institutions are structured and function can help one understand better:

- how development policies and operating rules might be altered when they become deficient (or when they are perceived to be deficient), and
- who has the authority to change such rules, and in what ways.

To achieve greater levels of sustainability, the rules set by institutions, eventually affecting such operational issues such as reservoir operation and water allocation, will no doubt have to change over time. This change must take place in many water resources institutions at various local, regional and national levels throughout the world. In response to demands from their constituents, the institutions themselves must be involved in this change. It is through these institutions that actions are taken to achieve a more sustainable use of natural resources and an improved quality of life. No real change toward more sustainable systems will occur without corresponding changes in the institutions that can implement those changes. And no real changes within or among institutions will occur unless it confronts the accumulation and use of power and how it operates in our everyday lives (Flyvbjerg, 1996).

SOME THINGS TO REMEMBER

- The social components of water resource systems are often the most challenging because they involve changing the way individuals think and behave. However, society is adaptive, and its institutions and individuals can change over time in ways that allow society to work better.
- The consideration of sustainability criteria along with the more common economic, environmental and social criteria may point to the need to change how we develop and use our resources. Any process involving change will require that we change our institutions – the rules under which we function.
- The notion of institutional levels makes it possible to address questions of power and collective action.
- To understand fully the boundaries of relevant institutions, the planners must be very careful to be sure they understand the 'real world'. They will also want to know, and are usually interested in how institutions function under stress, or under pressures for change from individuals within and outside the institution.
- Understanding how institutions are structured and function can help one understand better how development policies and operating rules might be altered when they become deficient, and who has the authority to change such rules, and in what ways.

9 Sustainability and modeling technology

Computer-based modeling technology plays a critical role in identifying and comparing or assessing the relative sustainability of proposed water resources management plans and policies. Without simulation models it would be difficult to predict the expected future impacts of any proposed plan and management policy. The planning, design, management, and operation of increasingly sustainable water resource systems depends on the modeling technology, and its continual development and improvement. In this chapter we examine some issues related to modeling and water resource system sustainability.

Interactive computer-based predictive models linked to data base management tools are a primary means available today for identifying and evaluating alternative water resource system designs and management or operating policies. Plan or policy evaluation involves predicting and evaluating economic, environmental, ecological, and social impacts. Many of these predictions are uncertain since many of the assumptions on which they are based are uncertain. Model outputs should be compared to actual events that take place over time. If the predictions are not as expected, either due to changes in objectives, to forecast errors, or to inadequate knowledge and data and/or models upon which the previous decisions were made, the modeling can be modified or the decision-making process can be started again. Models can help identify the impacts of the uncertainty on the predictions through sensitivity analyses – by being able to change the model input data and even the model equations and assumptions, running the modified models containing these changed assumptions, and evaluating the results. Modeling technology is constantly changing in response to changing information needs, changing computer technology and improved knowledge of the systems being modeled.

ADVANCES IN PLANNING TECHNIQUES

To identify those water resource system designs and management policies that are more sustainable than others requires being able to predict, quantify and compare the effects or impacts of those systems with respect to different perspectives or system performance criteria. Extensive research over the past decades has provided many tools available for such prediction efforts. These tools make it possible to perform integrated broad-based analyses of alternative water resource development plans and policies. No longer need one focus on single-sectorial single-purpose planning such as for water supply or hydropower or irrigation. What used to be the sole domain of agricultural or civil (hydraulic and sanitary) engineers is now the concern of a much broader and wider range of professionals as well as a much more informed public. All involved stakeholders can be aided by the use of computer-based interactive systems specifically developed to perform more comprehensive multi-sector, multi-purpose, multi-objective water resources planning and management studies.

Comprehensive modeling of complex, multi-component, water resources systems has been considerably enhanced by developments that permit various mathematical 'objects' to be linked to data bases and operated through computer graphic user interfaces. Mathematical optimization and simulation modeling techniques often together with geographical information system capabilities are now widely incorporated within numerous interactive software programs developed to study water resources planning and management issues. Many such issues can be analyzed sufficiently well for many decisions using comparatively simple conceptual hydrological models. More complex hydraulic models, often necessary for more detailed engineering design, are also finding increased applications, in part because of their increasing ease of interactive use made possible by advances in computer technology.

Hydrologic and hydraulic modeling using mathematics and computers have replaced to a large extent the need for more expensive physical modeling. Computer-based models of water systems are increasingly including morphological, biological and chemical processes which permit the study of pollution problems in rivers, ground water and other water bodies and their impacts on the aquatic ecology.

Predictions of multiple impacts are usually expressed in lengthy time series. Special statistical procedures have been developed, e.g., for time-series analysis of precipitation and stream-flow quantity and quality data. Statistical procedures are also available for estimating extreme events and their probabilities and for predicting the uncertainty associated with groundwater flows and qualities.

Models can be calibrated and verified with respect to past events, but this does not guarantee the accuracy of predicted future events. Assumptions concerning the future, the parameter values of models, and even the models themselves, are based on judgments. Evaluation processes involve comparisons of numerous physical, environmental and social impacts over time and space associated with different assumptions. These comparisons must be made based on multiple and often conflicting objectives or goals. Finally, with some understanding of all this uncertain and possibly conflicting information, a choice – a decision – must be made. If the decision is robust with respect to future uncertainties, it might also be sustainable.

Decision support systems

Repeated use of computer-based models and data bases under different assumptions, whether for system design or for real-time management (where suggested operating decisions are obtained periodically by resolving the models with updated data) is considerably facilitated through the use of interactive, menu-driven, graphics-based user interfaces. Such interfaces give users interactive control over data entry, editing and display and over the particular operations being performed by the computer. This interaction and display capability facilitates user understanding of both the input and output. Such interactive computer programs are typically called *decision support systems* (DSSs).

Decision support systems are interactive computer-based information providers. They, like their underlying models and data management components, do not make decisions. They merely provide information to those who need it or who can potentially benefit from it. Decision support systems for water resources planning and management provide a means of examining the numerous considerations involved when attempting to design and manage increasingly sustainable water resource systems. These considerations will include the extent to which the water resource systems can contribute effectively and equitably to the economic and social welfare of its users and at the same time enhance the environment and protect the supporting ecosystems.

Decision support systems can also support an adaptive, real-time planning and management approach, in which the decisions as well as the decision support systems are continually updated and improved over time. In multi-reservoir systems real-time management may be the only way we have today of most efficiently managing in a coordinated way the flows and storage of water, the generation of hydropower, and the allocation of water to the multiple users throughout a basin. The computer modeling technology is available, but the institutional arrangements that may permit its use often are not.

The analysis tools included within any DSS will depend on the problem being addressed and on the skills and judgments of its developers. The tools can include expert systems (knowledge-based or intelligent decision systems that exploit artificial intelligence techniques to process rules and symbols, to draw conclusions principally through logical or plausible inference sequences, and provide users with an explanation of how the conclusions were reached), executive information systems (data bases, analysis tools and interfaces that address the needs of top managers), and group negotiation support systems (that support teams of cooperative and uncooperative decision-makers). In addition, DSS's can

- include optimization and simulation models that can find values of decision variables or system performance indicators given inputs and constraints,
- include geographic information systems (GIS) that permit analyses and map displays of spatial data,
- use genetic algorithms that can help in the calibration of physical and chemical process models as well as deter-

mining parameter values of system design and operating policies, and

- include neural networks that can learn to reproduce results of complex processes and hence serve as 'black boxes' for those processes.

Decision support systems can accept and present data in a wide variety of forms, ranging from tables and text, to static and dynamic time- or space-series graphs, to static and dynamic (animated) pictures and displays on maps, to sound and video, to virtual reality. What is included within a particular DSS and the equipment it runs on depends on the DSS development team and on the resources and requirements of the end users.

Decision support systems not only can serve as tools for analysis, but also as vehicles for communication, training, forecasting and experimentation. They are typically application and problem oriented rather than methodology oriented. They can serve as links between field experts and decision-makers, i.e., between science and social policy.

The building and use of DSSs for identifying sustainable water resource system development and management plans and policies is typically an adaptive process. Decisions will be made based on the best information available, but under considerable uncertainties concerning what will happen in the future. The impacts of these decisions should be monitored, and as necessary revised actions should be taken based on more current information and objectives. What follows are some examples of situations where DSSs have been, or are being, developed and applied to assist in that process. The technical details of the particular DSSs can be found elsewhere. Here the focus is on the issues being addressed and how the DSS technology may be able to help.

DSSs for sustainable water resources planning in the European Union

A major five-year program was initiated in 1992 in Europe to develop a DSS for integrated river basin management. The purpose of this DSS is to assist managers in coping with the complexities of multi-objective sustainable planning within imposed environmental, public acceptance and legal and administrative constraints. The project involved the collaboration of:

- Centro IDEA of the University of Bologna, Italy
- University College Cork, Ireland
- University of Newcastle upon Tyne, United Kingdom
- International Institute for Applied Systems Analysis, Austria

- Ansaldo Industria Spa., Italy
- Thames Water International, United Kingdom

The universities concentrated on the analytical techniques, the International Institute for Applied Systems Analysis assumed responsibility for integration and packaging, Ansaldo Industria and Thames Water International undertook the prototyping. These efforts were to be applied to the Arno River in Italy and to the Thames River in England. The latter went ahead, but the Rio Lerma in Mexico took the place of the Arno River (Jamieson & Fedra, 1996b). The original cost estimate for this research program was in the order of US $15 million. The project is ongoing.

The main aim of the program was to provide the water industry with a methodology for managing water resources (including rivers, aquifers, lakes, reservoirs, estuaries and coastal waters) in a sustainable manner (Fedra, Weigkricht & Winkelbawer, 1993). The system was to be customized for each application and should allow managers to address a wide range of planning issues connected with water quantity and quality. It is designed to address problems associated with:

- forecasting future water-related requirements,
- assessing the environmental impact of socio-economic proposals,
- determining the limits of sustainable development,
- deciding what, where and when new resources should be developed,
- formulating strategies for river and groundwater clean-up programs, and
- evaluating the effect and hence the cost of implementing new environmental legislation.

No single software program would be capable of coping with the complexity of these issues in every river basin. Therefore, the approach adopted has been to create a flexible, modular package based on a 'tool kit' which allows users to select the components required to manage their particular catchment. The system is designed to integrate geographic information systems (GIS), data base management systems, modeling capability, optimization techniques and expert systems.

Provided the new components comply with generic interface definitions, there are no limitations on the capability to replace definitions or to replace existing or add new component models. Indeed, it is likely that components such as estuaries will eventually be represented by more than one model that users can select, depending on the degree of sophistication required and the data available. Social elements and some environmental considerations are introduced using expert systems, rule-based inference and qualitative reasoning. Other tools coupled to the modeling techniques will provide estimates of the direct economic costs and benefits of alternative engineering proposals. The intention is to create a system in which the mechanics of linking one component with another is largely transparent to the user. Although time and effort will be required to customize the system to a particular river basin, its use will be as simplified as possible.

Communication between the computer models and humans is by means of a graphics-based interface which makes extensive use of hypertext. If the user is familiar with the system, he or she will be able to call the appropriate component model directly. If not, the problem can be described and the expert system asked to select the appropriate component model. The philosophy adopted is that of a completely open, modular system with different degrees and mechanisms of coupling at various levels of integration. The core of the system is a set of conventions, standards, tools and a common language for problem representation, rather than any specific software system.

WaterWare is the name that has been given to the model resulting from this project. It encompasses a set of standards to ensure compatibility and consistency among any user-selected modules that may be applicable to the specific water resource system. The main program coordinates the individual tasks and provides access through a menu of options by means of either a single screen with triggers to individual applications or components, or an expert system for coordinating individual components in a problem-specific manner. The GIS stores and displays all geocoded information. The data base manager provides access to non-spatial data. The optimization, simulation, and/or expert system components provide the predictive analytical capability. The set of pre- and post-processors support the input, editing and output of data and its visualization. The user interface and various utility functions complete the set of model components (Jamieson & Fedra, 1996a; Fedra & Jamieson, 1996).

As evidence that the development of such integrated generic water resource planning DSSs is not a trivial task, this team of talented experts with considerable funding have taken over 5 years just to develop the basic framework of WaterWare. If they succeed to meet the goals they set for themselves, their product should provide an improved means of examining sustainability issues in water resources development using a variety of models (as needed for the particular basins and issues of interest) in an integrated manner. Traditionally these models were used separately without substantial feedback and interaction.

DSSs for sustainable groundwater management in eastern Germany

In the Lusatian district of Germany lignite (brown coal) open-pit mining began in the 18th century. The steady increase in mining until 1989 was interrupted only by World War II and a period in the 1960s. At that time a change in the energy policy from lignite to oil and gas was planned. The oil crisis stopped this planning process and mining increased further around 1970. In 1989 about 183 million tons of lignite were produced (Kaden, 1994).

Lignite coal seams in the Lusatian region are commonly embedded in aquifers. Hence it is necessary to de-water these mines by pumping ground water prior to and throughout the mining operation. In loose rocks the surrounding aquifers also have to be drained to maintain the stability of the slopes and the bottom of the open-pit mines. For that purpose about 1,200 million m^3/year of ground water have had to be pumped for mine drainage.

Following the German reunification in 1990 the economic importance of lignite sharply decreased. In 1992 only about 85 million tons were produced. This is an excellent example of how strongly the character and degree of environmental impacts can bring about changes in social and economic development – and vice versa.

The environmental and ecological impacts as well as the economic consequences of open-pit lignite mining in this region of Germany have been substantial. In sandy aquifers in particular, the mine drainage leads to the formation of large cone-shaped groundwater depressions. As a result small rivers dry up and large ones lose part of their flows due to the infiltration of water from them into the depression zone. Water supply and agricultural production also suffer from the lowering of the groundwater table. Wells for water supply can go dry and have to be replaced. Ecological conditions are also altered by lowering the groundwater level. Wetlands and park landscapes are endangered by the lowering of the groundwater table. Surface water ecosystems certainly suffer from water depletion and pollution. Additionally, water quality is affected by mine drainage due to the oxidation of ferrous minerals (e.g., pyrite) in the de-watered underground layers. In the cone of depression the overburden is aerated. With the natural groundwater recharge the oxidation products are flushed out, and the percolated water becomes very acidic. The same effect occurs during the groundwater rise after the mines are closed. Finally serious contamination risks for groundwater are incurred by the disposal of liquid and solid wastes in the mining region. These wastes together with very toxic military wastes (a typical problem, especially in the former East

Germany) may be washed out with rising groundwater tables.

These examples indicate that large scale mining can cause major impacts on the regional water resources systems, on the regional environment as well as on the economy in the region. Can one make development in the mining region sustainable? There are no simple answers. And this is not only because of economic and social consequences (such as unemployment in the region). It is also due to environmental consequences.

The main river in the Lusatian region is the Spree, flowing through an important wetland, the Spreewald (a UNESCO Biosphere Reserve) and finally through Berlin, the old and new German capital. The water supply of Berlin strongly depends on the Spree River (mainly from bank-filtration). The runoff of the Spree depends to a large extent on the flow augmentation due to mine drainage. If groundwater pumping were to cease, the natural flow in the Spree would largely infiltrate into the ground due to the large cone-shaped depression and the Spree would practically go dry. The total volume of the cone of depression amounts to over 10,000 million m^3. About 20 years would be required to fill this cone, using the total runoff of the Spree. This is unacceptable. Consequently, strategies have to be found to achieve sustainable development – at least for the remediation period which must be considerably longer than 20 years (Kaden & Schramm, 1993).

Groundwater pollution due to the excessive use of fertilizers and manure for increasing agricultural production is another case where the principles of sustainable development have been neglected. This not only happened under the pressing economic conditions of the former GDR, but also in the highly developed western German economy. While this has been partly due to a lack of knowledge, it has also been due to political, economic and social constraints.

A third example is groundwater pollution due to hazardous wastes. This again is an important problem where the concept of sustainability has been neglected. In Germany about 100,000 hazardous waste sites have been identified. Over 10 percent are expected to endanger groundwater resources. The financial requirements for remediation measures greatly exceed the funds available. Hence, in Berlin only those waste disposal sites that directly affect water supply are considered for remediation (i.e., only those that are located in water protection zones). Because of the financial restrictions, all other waste disposal sites have to be neglected. Consequently current generations and their descendants are faced with a large number of chemical 'time-bombs' in the ground. Is this sustainable? The mini-

mum to be done now is to avoid future hazardous wastes (already a very difficult task) and to reduce the most pressing hazardous waste risks in a way that efficiently and effectively leads to a more sustainable future. The solution of such problems cannot be effectively achieved without appropriate modeling, i.e., decision support.

These examples illustrate that in the past due to the lack of knowledge, as well as to various political, economic, technical and social constraints, the basic principles of sustainability with regard to groundwater resources have been neglected. At the moment decision support is needed to identify and evaluate alternative remediation strategies. Clearly in the future, sustainability must be considered in the project planning phase, not just in the cleanup stage. The law regarding environmental impact assessment in Germany (UVPG, 1990; German Federal Ministry for the Environment, 1994) should help to achieve this. But lack of knowledge, economic constraints and various short-term economic goals of the people and their political representatives will probably continue to cause problems in the future.

Water management related to mining is a typical example of changing objectives over time. For example, by the beginning of this century in the industrial region of the 'Ruhrgebiet', the Emscher, some 400 km of watercourses, including all tributaries, had been transformed into an open sewerage system (Schmid, 1993). Currently a large reconstruction program is underway. The former channeled system is being re-naturalized, i.e., ecologically remodeled. Did our early decision-makers and engineers do wrong as seen through our eyes? Are we doing it right now, or will our descendants view the world differently and line the river banks with concrete again?

The water management problems in the Lusatian lignite mining region have been examined with several successive decision support systems, beginning with groundwater modeling for the design of mine drainage systems in the 1970s followed by the development of a prototype decision support system for the mining region in the 1980s and, finally, by a water management study during the 1990s.

With the increase of lignite mining in the 1970s and the corresponding increase of mine drainage required, the effective design of the mine drainage system (well galleries) became important. Appropriate groundwater flow models had to be developed, considering the specifics of the hydrogeological setting (a multi-layer aquifer system) as well as the typical boundary conditions (Kaden, Hänel & Seidel,1976). Such models are classic simulation ones, i.e., they are used to predict the groundwater development in space and time depending on the given technological assumptions.

Effective technical solutions and management strategies have to be found by trial and error within a scenario analysis. The model, TAFEGA (Kaden, Hänel & Seidel, 1976), was used for many years to successfully design the drainage systems in the Lusatian lignite region. Sustainability objectives were not considered explicitly. The DSS was applied to minimize mine drainage and maximize mining safety. The client was the mining company. At that time environmental aspects were not of practical interest. Efficient lignite production was the basic objective.

In the 1980s environmental aspects received more attention in mining regions, although their impact was highly restricted because of the continuing priority of lignite production. An expression of this was the development of the decision support system MINE. The DSS MINE was developed at the International Institute for Applied Systems Analysis (IIASA) in cooperation with different institutes of the former GDR (Orlovski, Kaden & van Walsum, 1986). The general objective was to support the development of rational long-term water policies which could reduce impacts of mine drainage on the natural water resources system as well as on the environment, and on the socio-economic development in the region. Both groundwater and surface water systems had to be taken into account.

The dynamic system (time horizon of 50 years) was modeled using a hierarchical approach. The model hierarchy depended on the step-size and on the available mathematical models. In the given case a two-level model system was developed consisting of a first level *Planning Model* and a second level *Management Model*. The Planning Model was designed for screening of principal management/technological decisions by means of a dynamic multi-criteria analysis for a relatively small number of planning periods. The Planning Model was developed for the analysis of rational strategies for long-term development. These strategies were selected considering a number of *criteria* (objectives) chosen from a set of *indicators*, e.g., cost of water supply, cost of mine drainage, satisfaction of water demand and environmental requirements. Monte Carlo simulation was used in testing the feasibility of the identified strategies.

The DSS MINE was highly interactive. The user could vary data by an intelligent spread-sheet type of menu. Additionally, computer graphics (color) was used for visualization of results.

In 1984/1985 these interactive techniques were sophisticated. Based on today's technology, they seem to have been rather primitive. Nevertheless, in a retrospective view, the model system DSS MINE satisfied to a high degree the requirements of a DSS for sustainable development,

although this term was not explicitly considered at that time. The system and its basic approach could be of value even today.

In 1988 the economic crisis of the GDR accelerated (and resulted later in) the demise of the GDR. Consequently there was decreasing interest in long-term strategies. Decisions were made on operational (frequently irrational) bases. That is why there is no real proof that the system could have been used successfully in practice. What had not been considered in the model was of course the drastic economic change that occurred in 1990. Nevertheless, the model could now be used to analyze consequences of these changes.

Under the control of the German Federal Environmental Agency (Umweltbundesamt) a project was started in 1992 to develop strategies for the remediation of the lignite mining region (Dornier, 1993). A private firm (WASY) was asked to contribute with water management studies with special regard to the surface water system. A revised simulation model (similar to the management model within the DSS MINE) for the whole Spree River basin was developed. The Lusatian lignite district is one part of it.

The groundwater component (called FEFLOW) required three-dimensional (3-D) modeling of groundwater flow and contaminant transport (Diersch, 1993; Diersch, & Kaden 1994). Such processes play an important role in long-term aspects of sustainable groundwater systems. FEFLOW is an interactive, graphics-based hierarchically structured groundwater modeling system for 2-D or fully 3-D, fluid-density coupled or uncoupled transient flow and chemical species transport in subsurface water resources. It contains graphical editors and mesh generators for the geometric design and discretization of possibly complex study areas, as shown in Figure 9.1). It is also capable of more general computational techniques to solve a wide class of subsurface flow and mass transport problems characterized by flexibility and robustness, and many additional graphical tools to manage the entire solution process and model data.

Two aspects of ground water clearly relate to sustainable development – the design of water protection zones and the establishment of proper monitoring systems. Both depend on the hydrogeological setting and forecasts of future events and system performance. FEFLOW offers a tool to do this.

Decision Support Systems such as FEFLOW will always be based on uncertain parameters and on partly unknown physics. With increasing knowledge and improved tools, new insights in processes and problems may be gained. Assumptions and decisions made now, e.g., for protection zones, may prove to be incorrect later. If so, adjustments can be made. Impacts, constraints and the objectives will

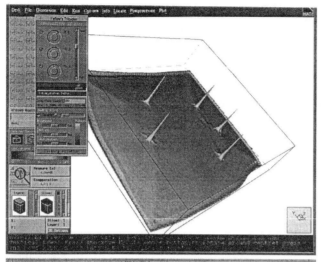

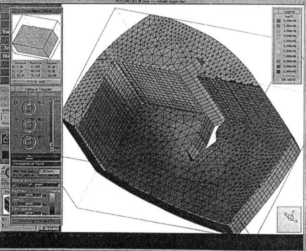

Figure 9.1 Sample displays of three-dimensional groundwater analyses performed using the FEFLOW decision support system.

also change during the next decades, as will our knowledge and tools.

DSSs for predicting economic, environmental and ecological impacts of water resources development

Since water resource management policies impact the economy and environmental and aquatic ecosystems, any assessment of their sustainability must include the future impacts. Any proposal for changing the current land and water use patterns in any local watershed, or entire river basin, should be judged in part based on the predicted economic, environmental and ecological impacts. DSSs are needed to provide local, regional and national planning agencies and decision-makers with the capability to make such predictions and

evaluations. Considerable research is currently underway in North America to develop this capability.

In addition to addressing local development issues, DSSs are needed to address the restoration of aquatic ecosystems in developing areas that have been degraded due to past and current land and water allocation and management practices. How can these ecosystems be restored? What will be the likely socio-economic impacts of any measures taken to improve the quality and vitality of such ecosystems? Restoring and maintaining the integrity of freshwater resources and aquatic life increasingly is being considered worthy of attention at local as well as national levels. Ecosystem restoration has become a major component of sustainable development policies in many countries throughout the world.

Fully restoring natural hydrologic cycles and ecosystems of watersheds impacted by humans is usually not feasible or desirable because of societal needs for, and economic and cultural benefits derived from, human activities on these watersheds. Nevertheless, we are learning that it is both feasible and practical to minimize disruptions of water movements and their negative consequences (e.g., non-point-source pollution) by maximizing the benefits derived from land and water management policies within a watershed.

A consortium of government agencies, universities and private firms in Europe, North America and Australia have undertaken the task to develop such a DSS (Taylor and Behrens, 1996). Their goal is to improve the capability of predicting a variety of economic, hydrologic, water quality, and ecosystem health indicators associated with alternative land and water management and use decisions. The methodology will include an interactive, object-oriented, data-driven, DSS designed to provide this information for, and at the levels of detail desired by, managers responsible for planning, managing and approving proposed land and water use patterns and conservation activities. Like WaterWare the DSS will integrate surface and ground water, water quantity and quality, aquatic ecosystems and human socio-economic (including land-use) activities and processes and will enable the display of these impact predictions, with their accompanying uncertainties. The methodology will enable one to estimate the sensitivity of the predicted impacts on the various ecological, environmental, hydrological and socio-economic data. It will be developed for implementation on multiple computers over Internet in collaboration with potential users. The intended users are planners, developers, environmental and other special interest groups, and consultants involved in making and/or influencing land-use and water management decisions as well as university groups studying these issues. The DSS is to be a 'shell' into which the particular water resource system of interest can be drawn in as a node-link diagram; the data needed for each node and link can be entered via the keyboard or graphically. The defined model can then be operated and the results displayed in multiple ways. The DSS will be designed to be used in water resource system model building exercises involving all interested stakeholders. Such exercises can help these stakeholders develop a shared vision about how their system works and how it should be managed.

This DSS will be building upon the object-oriented, data-driven, simulation models already developed for predicting the impacts of alternative water quantity and quality management policies on the flows, storage volumes, and constituent concentrations in aquifers, wetlands, streams, rivers, lakes and reservoirs in interdependent surface and groundwater systems (examples are numerous, and include the AQUATOOL model (Andreu *et al.*, 1996), IRAS (Loucks, French & Taylor, 1995), Delft Hydraulic Lab's RIBASIM, and PCRSS (Behrens & Reinink, 1994). This object-oriented modeling approach will be expanded to include GIS-based objects that define the spatial and temporal distributions of water quantity and its constituents (e.g., sediment and nutrients) in the runoff from watershed lands. It will be linked to demographic/economic models for predicting the probabilities of possible changes in land use over time driven by economic growth and various land and water use policy decisions. These land-use decisions in turn may alter the quantity and quality of the runoff and hence have a long-term impact on aquatic ecosystems.

The overall structure of the DSS will include four interdependent components combined within a synthesis component (Figure 9.2). Module 1 predicts the probable changes in local land use that are likely to occur over time in response to economic and population growth projections as well as in response to current or proposed land-use and water management decisions. These changes may impact surface and groundwater quality, and hence the aquatic ecosystems themselves. Modules 2 and 3 simulate the rainfall-runoff and water flows, temperatures, oxygen concentrations, and sediment and nutrient loadings over time. Module 4 defines the relationships between environmental parameter values (gradients) and ecological performance or integrity. To predict any indicator of ecosystem type and health, the relationships between sets of time- and space-varying values (or gradients) of environmental parameters and ecosystem type or health must be identified. Calibration must take into consideration the adaptive and evolving characteristics of ecosystems. Module 5 is the interactive interface for data input, editing and analysis, for controlling the operation of each of the four component modules, and for creating various displays of input data and results on a microcomputer.

Input data to this DSS must include relationships between environmental variable values (or gradients) and ecosystem health. One approach for doing this is through the use of

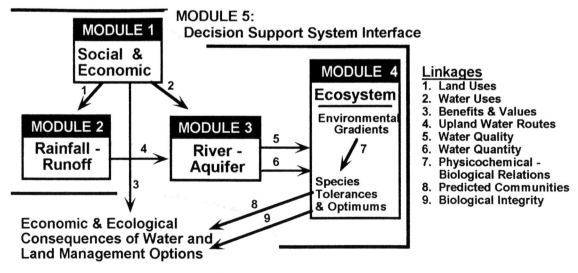

Figure 9.2 Decision suppport system modules and links.

logistic regression equations of a bell-shaped curve for estimating the probability, p, of the presence of a particular type of ecosystem:

$$p = \exp[f(x)]/\{1 + \exp[f(x)]\}$$
$$\text{where } f(x) = ax^2 + bx + c \qquad (9.1)$$

This is a calibrated regression equation of some environmental variable, vector or index, x. The probability, p, has a maximum value at some (optimum) value of the index, x, and a symmetric spread or tolerance on either side of the optimum value of x. Equations of this form can be extended to include multiple variables, vectors or indices. Such equations form the basis of the ICHORS computer program used in parts of Western Europe (Barendregt & Nieuwenhuis, 1993; Hooghart & Posthumus, 1993).

The gradient-based methods described above are only part of the ecosystem modeling exercise. The other part involves model calibration and how such models can be made to evolve and adapt over time. There is a growing recognition that many natural (including social) processes we wish to predict cannot easily be modeled deterministically. What this means is that living systems change. Hence each ecosystem 'object' must have a self-learning, adapting capability, much like we humans have as we adapt, or evolve, to changes in our environment and to changes within ourselves. Models of ecosystems, including ourselves, change over time. Not only may the parameter values change, but so may the model equations. This emergent property of modeling is inherent in genetic algorithms (for finding the 'best' parameter values based on observed data) and genetic programming (for finding the 'best' fitting model or equations based on observed data). To the extent that ecosystem objects can learn and adapt and evolve, such processes should be modeled. This might be possible using genetic-based modeling procedures.

The pursuit of evolutionary-algorithm-based, but still relatively simple-to-use, ecological models has been undertaken by researchers (e.g., at University of Adelaide, Cornell University, University of Illinois, University of Maryland, University of Texas, TU Delft, International Institute for Hydraulic and Environmental Engineering, (IHE) Delft, and the Danish Hydraulics Institute (DHI), among many others. The integration of individual ecosystem communities within the real-world dynamics of large ensembles of communities, effectively accounting for adaptation, change and evolution (e.g., from one species group to another), remains a significant research challenge (Fitz et al., 1996; Hooghart & Posthumus, 1993; Maxwell & Costanza, 1994; Naiman et al., 1995).

APPROPRIATE MODELING TECHNOLOGY FOR SUSTAINABILITY

Whatever modeling technology is developed and implemented to study a particular water resource system, it cannot address sustainability issues unless it addresses or simulates the variables of concern to those who will be affected by its management. It must not only focus on what is economically of interest, but also on what is culturally or socially important. This should be emphasized. *The value of a particular predictive technology lies not only in its economic viability and its technical soundness but also in its adaptation to the local institutional and cultural environment.* Engineers and computer scientists involved in coding DSSs are typically not well trained in identifying the latter. Hence the need for the participation of individuals involved in the planning and decision-making process during the development of DSSs that are to address sustainability issues. Through them there can be a greater understanding of how to handle a variety of legal, institutional and political issues. Legal issues and constraints include laws, court decrees and treaties. Institutional procedures and historical precedents include practices adopted some time ago that are now part of the institutional culture. Politics often involve existing and potential future alignments and conflicts of interest among different stakeholders.

Although not specifically addressing sustainability issues, Reitsma (1996) outlines some of the experiences he and his DSS development group (at the University of Colorado in the USA) have had as they have worked with two major clients involved in regional water resources planning and management in the USA. These and other actual case studies of the use of DSSs (e.g., Loucks & da Costa, 1991; Simonovic, 1996) all point to the major challenge to all who work toward developing improved models for studying water resource system management and sustainability issues: the effective integration of these DSSs into the institutional and social infrastructure of the organizations responsible for water resources management, now and into the future.

Institutional and social infrastructures are characterized by the organizations people in a community create for themselves. Through these organizations, people identify common development needs and objectives, priorities and resources. Through these organizations, people also mediate between competing interests; establish committees, task forces and cooperatives; and monitor and oversee development and operations of water resource systems.

Ideally, engineers must work within these infrastructures to ensure that their proposed technologic solutions are appropriate for, and compatible with, the values of the people affected. Too often, however, the education of engineers and other professionals promotes a feeling in which they are assumed to have all the knowledge needed for decision-making. Hence, information about the impact of a given technology is not always shared. This can lead to the selection of technologies that may be appropriate for the economic and engineering objectives as perceived by the experts, but are inappropriate to the community that is to be served by these technologies. An appropriate modeling technology for sustainability may differ from that designed to meet traditional economic growth criteria. Economists often use models tracking the amount or quantity of goods and services we consume as one measure of our standard of living or well-being. This leads to policies favoring technologies that maximize consumption. Alternatively, consumption can be viewed as a means to improved human well-being. But well-being involves much more than consumption of goods and services. The aim should be to maximize well-being with the minimum of consumption. Other attributes of well-being that are among the conditions required for achieving sustainability include:

- compatibility of technology with nature and the physical, biological, ecological, social and cultural, political and sensual (including visual) environments,
- use, preservation and protection of renewable resources,
- equitable allocation and sharing of all benefits and costs,
- emphasis on service as well as profit and financial feasibility,
- enhancement of local social institutions and cultural traditions,
- public involvement in all aspects of planning and management and decision-making.

THINGS TO REMEMBER AND CONSIDER WHEN DEALING WITH SUSTAINABILITY MODELING

- Without simulation models it would be difficult to predict the expected future impacts of any proposed plan and management policy.
- All involved stakeholders can be aided by the use of computer-based interactive systems specifically developed to perform more comprehensive multi-sector, multi-purpose, multi-objective water resources planning and management studies.
- While models can be calibrated and verified with respect to past events, this does not guarantee the accuracy of predicted future events – the assumptions made concerning the future could very well be completely wrong.
- Decision support systems for water resources planning and management can provide a means of examining the numerous considerations involved when attempting to design and manage increasingly sustainable water resource systems. Decision suppport systems can also suppport an adaptive, real-time planning and management approach, in which the decisions as well as the decision support systems are continually updated and improved over time.
- Whatever modeling technology is developed and implemented to study a particular water resource system, it cannot address sustainability issues unless it addresses or simulates the variables of concern to those who will be affected by its management. Thus, the value of a particular predictive technology lies not only in its economic viability and its technical soundness but also in its adaptation to the local institutional and cultural environment.

10 Sustainability, hydrologic risk and uncertainty

RISK, UNCERTAINTY AND SURPRISE

Sustainability criteria as discussed in earlier chapters of this document all involve projections of future conditions or impacts. Models, such as those mentioned in the previous chapter, are used to obtain these predictions. However anyone trying to predict what may happen or be observed in the future, especially the distant future, knows such predictions are at best risky, possibly uncertain, and subject to all sorts of surprises we cannot even imagine. Thus, all impact predictions entail some element of risk and uncertainty. Risks are usually measured by the probabilities that can be assigned to the likelihood of occurrence of an undesirable event (Ang & Tang, 1984). Uncertainty describes a situation where little is known about future impacts. Either no probabilities can be assigned to definite outcomes, or in fact the outcomes could be so novel that they cannot even be anticipated, i.e., they are subject to surprises.

Risk can be based on known or estimated data, and therefore can be insured against and treated like any other cost. However, uncertainty defies actuarial principles because of undefinable outcomes. Uncertainty is especially important in sustainability issues, where the amount of uncertainty becomes much larger than the measurable risk. The use of a single number (or expected value of risk) does not indicate the degree of variability or the range of values that might be expected. Additionally, it does not allow for individual perceptions of risk and uncertainty. A logical response to risk and uncertainty with respect to sustainable development is to *proceed with caution if the future cannot be perceived clearly*. The speed of advance into the future should be tailored to the distance over which the clarity of vision is acceptable.

In practice, the way risk and uncertainty are often included in any modeling study is through sensitivity analyses. Such analyses produce estimates of how various measures of system-performance depend on different system parameters or variables. Using optimistic and pessimistic estimates for different model parameter values that are uncertain can indicate which of those parameters will have the most pronounced effects on a system's performance. Although sensitivity analysis may not reflect the probability of occurrence of these two sets of estimated values, it can be useful for identifying which variables are the most important to the success or failure of a project, and which are not. Considerable literature has been written on the subject of risk and uncertainty in project appraisal.

The issue of future uncertainty plays an important role in environmental valuation and policy formation. *Option values* and *quasi-option values* are based on the existence of uncertainty. An option value is the premium that consumers are willing to pay to avoid the risk of some unfavorable event occurring in the future. A quasi-option value is the value of preserving options for future actions in the expectation that knowledge about what actions are best will grow over time. If a development takes place that causes irreversible environmental damage or loss of biodiversity, the opportunity to expand knowledge through scientific study of an undamaged environment or ecosystem is lost.

Sustainability analyses are complicated by many uncertainties concerning the future values of system performance indicators used to measure relative sustainability. These may include the identification of sources and amounts of future pollutants, the ultimate destination of particular emissions, the actual physical, biological and social impacts of these pollutants, the human valuation of the realized impacts,

108

the extent to which a particular policy response will have an impact on all these factors, and the actual costs and distribution of those costs that are the result of some policy decision.

Other sources of uncertainty linked with environmental issues include uncertainty about land tenure, which can lead to deforestation and unsustainable agricultural practices, and uncertainty about resource rights, which can accelerate the rate of depletion of a non-renewable resource. Policy-makers can address these issues by instituting land reforms and by designing appropriate taxation policies that return rents to public sources rather than to private agents.

The ways in which policy-makers address these uncertainties depend in part on their willingness to consider the expected (and uncertain) interests of future generations. In the absence of such interests, decision-makers may tend to follow policies that ignore (or at least heavily discount) costs to future generations, and that minimize costs to current generations at the expense of the future.

NATURAL AND HUMAN-CAUSED
HYDROLOGIC HAZARDS AND DISASTERS ——

Humans are subject to a multitude of natural water hazards, e.g., floods, rain-induced landslides and droughts. But these are relatively short term events with respect to the times considered for sustainability planning and evaluation. Hence the need to average out these relatively short time events when calculating system performance trends for sustainability analyses (for example as discussed in Chapter 4).

A hazard is defined as the potential for the occurrence of an extreme adverse event. When such an event actually happens, depending on its severity, its consequences may range from small economic damages to the loss of human lives. When these damages are extensive, the event is a called a disaster. The most frequent water-related disasters are those caused by extreme floods, floods that can develop relatively rapidly. In contrast, some disasters, such as those caused by droughts or water contamination, may develop over long periods of time. In these cases it is not always easy to identify just when a disaster begins, and hence when to take countermeasures to reduce the accompanying damages. Hence it is not always clear just how a future time series of system performance indicators should be grouped in order to calculate the average performance values over those successive groups.

Floods not only bring the obvious physical damages but can bring potential health hazards as well. Water supplies may become polluted from the overflowing of wastewater basins and septic tanks, sanitation facilities may become non-functional, and all water may become suspect and have to be boiled for drinking purposes. Obstruction and subsequent blocking of water courses by debris from landslides, by logjams and by volcanic debris can be major issues in mountainous areas – especially in highly sediment-yielding and erosive canyon-like terrain. A number of projects have been doomed by such obstructions that occurred in the life-span of those projects. Any geologically unstable feature should be accounted for in project planning and design before building and implementing a project. A good example is the Ganges Project (India) in which water resources development for energy is severely handicapped by landslide sediments feeding into the river.

People have also discovered that their water supplies can be threatened by disasters of their own making. These include those caused by human error in treatment plant operation, by long-term negligence of the system elements of plants in industrial production complexes or during transport of hazardous substances. Such non-natural disasters appear to be occurring more and more frequently. Consequences of such disasters, for example those due to chemical spills or nuclear accidents, may be far reaching – in both time and space. Sustainability implies a condition in which the threats to society, as well as the threats to water, are managed in such a way that people are prepared and able to cope with them when they occur, and that their frequency and severity are decreasing over time.

The beneficial role of a dam in a water resources development scheme is easily taken for granted. The reservoir behind the dam equalizes water supplies, except possibly in the case of a major drought. In such situations the manager tries to provide a reliable back-up system; for example, a groundwater supply is often an essential component of a reliable water supply system. However, failure of dams cannot be entirely ruled out. Apart from the probable disaster caused by a dam break event itself, such a failure can cause major disruptions in a water supply system. Many years can be required to replace the structure (if in fact it is even decided that it is desirable to do so) and for the reservoir to fill again to the level required for normal operation.

Well-known examples of loss of continuity, or complete failure of supply due to dam failures include the failure of the Vajont dam/reservoir (Italy) (in this case the catastrophic failure was due to a major rock-slide into the reservoir, causing a large wave to pass over the dam. The dam did not fail).

Others include the Teton dam (USA) which failed during the original filling of the reservoir, and the Vega de Tera (Spain) and Malpasset (France) dams, which were both lost.

As part of flood contingency action, dam legislation in many countries currently requires flood inundation maps, often based on hydrodynamic dam break models. Public awareness programs are implemented to ensure adequate evacuation drills and other procedures to be followed. Dam breaks caused by foundation failures as well as failures due to excessive runoff during strong rainstorms must be envisioned.

The prevention, management and control of disasters have a high priority in the achievement of sustainability. However, it is usually neither politically feasible nor economically possible to eliminate all potential hazards or to design all water resources systems to withstand any conceivable extreme event. One must (at a minimum) compare the cost of a disaster with the cost of preventing it. An event with extreme consequences that is also extremely rare will be much worse than a less consequential event that occurs more frequently. In either case the consequences can be quantified and compared by calculating the expected risk. The expected risk is defined as the average potential consequence of a disaster per unit time. It weighs the consequences of a disaster, such as loss of lives or monetary losses, according to their probabilities of occurrence. While such expected value calculations can be made, they are by no means the only relevant measures of risk of interest to planners and decision-makers.

The purpose of risk management is to make efficient and effective planning decisions that mitigate the effect of potential disasters. Risk management includes provisions to cope with a disaster in the unlikely event that it occurs. A primary purpose of disaster management would be to prevent, to the extent possible, both natural and man-caused disasters. A poorly managed disaster can disrupt the social structure of a region for many years. Natural disasters (and wars) have proven to be highly disruptive of life.

Risk management has been formalized in recent years. The first step in preparing for potential disasters is to identify the potential hazards existing in a region, and evaluate their probabilities of occurrence. Next, various sets of socially appropriate structural and non-structural measures for disaster prevention and mitigation should be identified, and their costs and expected benefits compared. Finally the best of these alternatives should be implemented, before the disaster occurs. These alternatives should consider the development of rules to operate an affected water resources system in such a way that it is resilient. After an extreme adverse event has occurred and caused a system to fail, the system should normally be able to revert back to a satisfactory state in a sufficiently short time and at an acceptable cost.

Within risk management, a number of methods can be implemented for disaster mitigation. These range from forecasting and warning systems to disaster resistant construction and disaster zoning. In the latter case, people are prevented from inhabiting regions of high potential disasters – like flood plains.

RISK-BASED DECISION-MAKING

'Risk-based decision-making' and 'risk-based approaches in decision-making' are terms frequently used to indicate that some systematic process is being used to deal with risk and other uncertainties when formulating policy options and assessing their various impacts and ramifications. Today an ever-increasing number of professionals and managers in industry, government and academia are devoting a larger portion of their time and resources to the task of improving their approach to and understanding of risk-based decision-making.

There are two fundamental reasons for the complexity of this subject. One is that decision-making under uncertainty literally encompasses every facet, dimension and aspect of our lives. It affects us at the personal, corporate and governmental level and during the planning, development, design, operation, and management phases. Uncertainty colors the decision-making process,

(1) whether it involves one or more parties,
(2) is constrained by economic or environmental considerations,
(3) is driven by socio-political or geographical forces,
(4) is directed by scientific or technological know-how, or
(5) is influenced by various power brokers and stakeholders.

Uncertainty is present particularly when the process attempts to answer, 'Who should decide on the acceptability of what risk, for whom, in what terms and why?' The second reason risk-based decision-making is complex is that it is cross-disciplinary, leading to the development of diverse approaches of varying reliability. Some methods, which on occasion produce fallacious results and conclusions, have become entrenched and appear difficult to change.

To ensure sustainability, considerations of risk must be an integral part of the overall management of a water resource system. This involves risk management. The term 'management' may vary in meaning according to the discipline and/

or the context involved. This is true also of 'risk' which is often defined as a measure of the probability and severity of adverse effects (Lowrance, 1976). Risk management is commonly distinguished from risk assessment, even though some use the term risk management to encompass the entire process of risk assessment and management. In risk assessment the analyst often attempts to answer the following three questions (Kaplan & Garrick, 1981):

- What could go wrong?
- What is the likelihood that it will go wrong?
- What would be the consequences?

Answers to these questions help risk analysts identify, measure, quantify and evaluate risks and their consequences and impacts. Risk management builds on the risk assessment process by seeking answers to a second set of questions (Haimes, 1992):

- What can be done, i.e., what options are available?
- What are the associated tradeoffs in terms of all costs, benefits and risks?
- What are the impacts of current management decisions on future options?

Only when these questions are addressed in the broader context of management, where all options and their associated tradeoffs are considered within the decision-making organizational structure, can improved sustainability be realized. Indeed, evaluating the tradeoffs among all important and relevant system objectives in terms of costs, benefits and risks cannot be done seriously and meaningfully in isolation from the overall organization's broader resource allocation perspectives.

Sustainable management policies must thus incorporate and address risk management within a holistic and all-encompassing framework, one that addresses all relevant resource allocations and other related management issues. A total management approach must include risk management as part of the overall system management of possible system failures. Of equal importance is the total involvement in risk assessment and management by everyone concerned.

In a society that continually adjusts to the risks of everyday life, a simple yet fundamental truth should be understood. In the face of such unforeseen calamities as floods, droughts and structural failures, it is important that the management of these 'extreme' (but unlikely, we hope) events be learned. Rather than expected risk, questions should be asked about expected maximum risk. In both a theoretical and practical sense, efforts should be focused on forming a more robust treatment of extreme events. Managers and decision-makers, of course, should be most concerned with

the risk associated with the specific case under consideration, and not with the likelihood of the average adverse outcomes that may result from various risk situations. In this sense, the expected value of risk, which until recently has dominated most risk analysis in the field, is not only inadequate, but can lead to fallacious results and interpretations.

In general people are not risk-neutral. They are often more concerned with low-probability catastrophic events than with more frequently occurring but less severe accidents. In some cases, a slight increase in the cost of modifying a structure might have a very small effect on the unconditional expected risk (the commonly used business-as-usual measure of risk), but would make a significant difference to the conditional expected catastrophic risk. Consequently, the expected catastrophic risk can be of significant value in many multi-objective risk problems.

Thus, unconditional expected value measures, while possibly adequate for some cases, are totally inadequate for questions of sustainable development and can easily lead to fallacious interpretations and policies if used as the sole criterion. To provide a clearer focus on the risk of extreme events (events of very low frequencies but with possibly catastrophic consequences), supplementary measures – such as the level associated with a probability of exceedence, or a conditional expected value – should be added to the operational principles of sustainable development. These measures can supplement and complement the unconditional expected value measure of risk.

The temporal dimension of risk is fundamentally critical in the consideration of sustainable development. It is not sufficient that tradeoffs between economic development and the risk associated with environmental protection be acceptable today. They must also be acceptable by (or at least deemed not damaging to) future generations. Indeed, executives and high-level managers already correctly consider strategic planning a precursor to an integral part of operational and tactical decisions. The virtue of strategic planning is that it aims at correlating the present with the future in terms of opportunities, options, policies, needs, consequences and impacts.

A component of strategic planning is sensitivity analysis (i.e., evaluating how levels of objective achievement or system output vary with changes in policy decisions and/or other system parameters). It is not uncommon for professionals to use sensitivity analysis as a surrogate for impact analysis, but the difference is fundamental. While sensitivity analysis is commonly performed separately from the decision-making process *per se*, impact analysis should be conducted as an integral part of the decision-making process.

Clearly, impact analysis must be one of the operational principles of sustainable development, which has at its core a concern that the welfare of future generations should not be jeopardized for shortsighted economic development that benefits only the present generation. Thus, impact analysis – evaluating the impacts of current decisions on future options – is yet another critical building block of sustainable development. Needless to say, good management must incorporate in its decision-making process the evaluation of all the risks (their likelihoods and associated adverse consequences) along with all the beneficial consequences resulting from the adoption of any policy.

FLOOD HAZARDS AND RISK MITIGATION —

Despite significant advances in flood forecasting and flood protection, increasing damages due to floods are a factor in both developing and developed regions. Witness, for example the Midwestern USA floods of 1993 that resulted in 38 human deaths, untold farm and wildlife animal death, and US $12–16 billion of damages, over half of which were agricultural. The damages to residential homes came from both the flooding and from the high groundwater tables or sewer backups. The cost of the emergency response and operations has been estimate at $6 billion. Other costs (business losses, etc.) have not been quantified (Galloway, 1994).

Another example is the flooding along the Mediterranean coast that occurred in 1982. In one day rainfall exceeded 600 mm and produced peak flow values on the Júcar River (Spain) of 9,500 m^3/s. Its mean flow is 40 m^3/s. That flood destroyed the Tous Dam, affected 300,000 persons, resulted in nine fatalities and caused economic losses of approximately $1 billion. In the 1982–92 decade economic losses due to flash floods in the Valencia (Spain) region alone amounted to more than three billion dollars. This is a significant economic and social impact for a region with four million inhabitants, of which about 700,000 persons were affected either at their residences or on their jobs.

Finally, consider Brazil. Stormwater drainage is becoming one of the most pressing problems facing cities in Brazil. For example, during 1979–89 more than 5,000 people died as a direct consequence of flooding while an estimated 3.6 million persons lost their homes. In Rio de Janeiro alone, in 1988, floods caused an estimated $1 billion in damages. In 1992, 35 people were killed in Belo Horizonte, as a consequence of heavy rainfall and mud slides.

Sustainable water resources development has to address the issue of managing and living with hydrologic extremes, extremes that are likely to increase in magnitude and frequency if global climate change research results are to be believed. In the discussion that follows, floods and droughts, and the measures usually taken to mitigate them, will be considered. This discussion is largely based on the research and writing of Valdez et al. (1995).

Flood impacts can include physical damages and the loss of life. The main difference between developed and developing regions is that in the former, flood losses are mainly monetary, while in the latter, loss of life is more predominant. In any benefit–cost analysis that includes the benefits of using flood plains for various economic activities, including the benefits of flooding itself (e.g., the refertilization of flood plains used for crop production), as well as the costs of flooding (e.g., the destruction of property and lives) would clearly show that some losses are to be expected if economic efficiency is the objective. Nevertheless, loss prevention measures are possible and should be a part of sustainable water resources planning.

Flood warning systems, even with minimal forecasts, can reduce the number of casualties. Recent flooding in Western Europe has shown that current forecasting systems are able to eliminate almost all human casualties. As an example, only one person was killed in The Netherlands and three in Germany in the 1995 floods in the Rhine Valley. At the same time, the economic losses were approximately US$ 6 billion and US$ 3 billion, respectively, in the two countries. Hence, warning systems can be a the key approach to protecting human life, even though their role in reducing economic damage is negligible. But human psychological damage can be high even when the death toll is low. Many people can lose all of their belongings while saving their lives. The 'beginning again' attitude prevails among those flood victims.

Flood forecasting capabilities have significantly improved in the last few decades. Ground-based Doppler radar is becoming more common in developed countries. Passive (and in the near future, active) space-borne sensors monitor areas in developed and developing regions and transmit the information in real time to weather and river forecast centers. These satellites can also be used as antennas to receive and transmit information from telemetric stations on the ground. This massive information can be processed by high speed computers and combined with databases that have geographic information systems (GIS) and other decision support systems (DSSs) to provide and update flood forecasts and warnings.

Seasonal to inter-annual climate predictions, in particular those related with El Niño-Southern Oscillation (ENSO), are becoming more and more reliable. These are currently producing forecasts up to 24 months in advance, successfully detecting the 1991 ENSO event almost a year in advance (Kerr, 1992). Several attempts to use these climatic forecasts in the production of operational hydrologic forecasts have been proposed in the literature and offer promising research opportunities.

Land-use zoning, a basic and fundamental preventive measure, is the main tool for land-use management of flood plains. Zoning assigns each land parcel a set of uses for which it is best suited according to its characteristics and flood hazards. Hence it serves to impede use of lands for activities that are most subject to damage in the event of a flood in the higher flood risk areas. In particular, zoning in high flood risk areas usually prohibits the building of residences or any other economic activities that could result in substantial losses due to flooding or that could contribute to added flood damages downstream.

Unfortunately, the risk of economic damage may be so high that both residents and authorities will opt to ignore it (perhaps assuming the government will help them out when flooding occurs). This is especially common in developing countries where the urban authorities have little enforcement power and growth is characterized by the abandonment of rural areas and explosive pressures on urban infrastructure capabilities. It is easy for the new residents, who are without roots in the new environment and who are ignorant of its flood characteristics, to locate in areas of high flood risk.

Strengthening of land-use and urban planning in many regions would prevent, with a minimal use of structural measures, situations that will require construction of hydraulic flood control structures after an area has been developed. Small easements today may substitute for expensive hydraulic structures in the future.

Flood proofing is the protection of individual residences and facilities, their sites and contents, against structural failure and the prevention or reduction of water entrance into the residence. Flood proofing requirements specified in building codes can help minimize the damages to buildings due to flooding. Extensive lists of various flood-proofing measures are provided in the literature. An extensive one developed by the US Water Resources Council (1973) is used in the US National Flood Insurance Program.

The US National Flood Insurance Program was created in 1968 and has impacted over 18,000 communities in the US. The basic standard is the definition of the areal extent of the 100-year flood (the flood that every year has a 1 percent probability of being exceeded). This standard defines the Special Flood Hazard Area as presented in Flood Insurance Rate Maps and participating communities are supposed to regulate land use in these areas. The participation of the communities in the insurance program and regulation of the uses of the flood plains are credited with mitigating some of the losses of the 1993 flood. Several recommendations to improve the program have been made, particularly in the enforcement of the required coverage, the supply of insurance and the waiting period for obtaining flood insurance, among others (Galloway, 1994; Kunreuther & White, 1995).

Water resources planning may also adopt a global hydraulic view of the flooding problem, especially for urban areas as proposed by Marco & Cayuela (1994). In their view, the development of a Flood Planned City is adapted to the topography and the alignments to the hydraulic behavior of flood flows, using the streets as supplementary canals and open spaces as storage areas. The possibility of flooding is considered at the initial planning level. Obviously, this can only be applied to new developments under low flood risk, and thus becomes relevant in developing regions where new urban areas are being built. The Flood Planned City concept also has a special design of public services such as power, drainage, telephones, gas, pavements, etc., for which new construction norms that include flood proofing have to be developed.

DROUGHT RISK MITIGATION

Drought is one of the most complex and least understood of all natural hazards, affecting more people than any other hazard (Wilhite, 1993). Droughts differ from floods, tropical cyclones, tornadoes, earthquakes and other natural hazards in several ways. The effects of drought typically accumulate over long periods of many months or years, without a clearly defined beginning and sometimes even an end.

A drought is broadly defined as a condition of widespread adverse economic, social and environmental impacts resulting from there being less water than is normally expected (Wilhite, 1993). It is the consequence of a natural reduction in the amount of precipitation received over an extended period of time in comparison to what is normally demanded. Drought characterization varies geographically, reflecting unique regional characteristics of climate, water resources, economic and social development and water management practices. Droughts, like floods, are the results of the natural variations in climate. Natural ecosystems tend to adapt themselves to extremes of floods and droughts. But societal and environmental vulnerability to droughts increase as population and economic growth intensifies the demands for limited water resources. The probability of climate change is also a consideration in assessing the increasing vulnerability of regions of the world to drought, and this is a substantial source of uncertainty with respect to sustainability analyses.

Most inhabited regions are vulnerable to droughts and the resulting economic, social and environmental consequences. While droughts are often associated with the arid regions of east Africa, west African Sahel, India, Australia and the Great Plains of North America, the more humid regions of the world also suffer from periods of below-normal precipitation. The occurrence of severe droughts world-wide during and following the extreme El Niño-Southern Oscillation (ENSO) event of 1982-83 demonstrated the vulnerability of all nations and captured the attention of the scientific community and governments world-wide (Wilhite, 1993). During the past decade, significant parts of Asia, Australia, Europe and North and South America have experienced extensive social, economic and environmental impacts from extended periods of severe drought.

Droughts can vary greatly in their spatial characteristics. They tend to be regional phenomena. Thus for smaller regions, the entire region may suffer the effects of a drought. In larger regions, perhaps only a portion of the region will experience drought. For example, in the USA the area affected by the severe drought of the 1930s never exceeded 65 percent of the country (Wilhite, 1993). In India, droughts of this century have rarely affected more than 50 percent of the country at once, with the exception of 1918–19 when 73 percent of India experienced drought conditions (Sinha, Kailasanathan & Vasistha, 1987).

The severity of impacts from any drought is related to its intensity, duration, timing and spatial coverage. Intensity refers to the degree of the precipitation shortfall and/or the impacts associated with that shortfall. 'Percent of normal precipitation' is the simplest of the various indices. Drought intensity is related to departures of actual precipitation from the long-term average and the duration of those deficits. Droughts typically result from large-scale anomalies in atmospheric circulation patterns that become established for long periods of time. Drought durations may vary from several months to many years. Impacts of drought also depend on the timing of precipitation shortfalls relative to crop-growing seasons and the timing of other ecosystem and human demands for water.

Drought impacts are less obvious and spread out over a larger geographical area than are impacts resulting from most other natural hazards. Damages to buildings and structures associated with floods and earthquakes are easier to assess than the diverse economic, ecological and environmental impacts of droughts. Consequently, development of emergency contingency plans, disaster relief efforts and damage assessments are typically more difficult for droughts than other disasters.

The impacts of drought depend upon human and ecosystem demands for water, available water resources management capabilities and practices as well as the meteorological and hydrological characteristics of the drought. Drought vulnerability is related to the sensitivity of various socio-economic and environmental systems to drought. The economic, environmental, social and political impacts of drought can be diverse, widespread and difficult to quantify. The effects can range from individual loss of income and rising prices to famine and international conflict. Extreme water and food shortages during droughts in lesser developed nations have caused public health problems, loss of life, population migrations, social unrest and civil strife. Droughts can cause conflicts between water uses in both developed and developing regions.

Agriculture drought impacts

Agriculture is particularly susceptible to drought. Although droughts affect both irrigated and dry-land agriculture, irrigation does lessen vulnerability. Drought impacts can include:

- losses in crop, livestock, dairy and timber production and associated effects on the economy,
- increases in costs of supplying water,
- erosion and other damage to land,
- forced reduction in foundation livestock,
- insect infestation and plant disease,
- forest and range fires,
- higher consumer prices for food and other products, and
- (in extreme cases) starvation and famine and associated social and political upheaval.

Municipal and industrial drought impacts and management

Droughts can have major effects on the supply of potable water for domestic and municipal use. Water supplies for municipalities as well as for irrigated agriculture are reduced during droughts due to limited streamflow, lowered reservoir levels and depletion of groundwater reserves. During droughts water supply agencies are faced with increasing demand combined with decreasing supply. They must balance meeting present demands with maintaining adequate reserves to continue to meet basic needs as the drought extends into the uncertain future.

The effect of drought on public water supplies necessitates cooperation between water users and local, regional and national public officials. But, since droughts are infrequent in many areas, water managers are faced with dealing with situations during droughts for which they typically have little or no past experience. For just this reason, the US Army Corps of Engineers have been actively engaged in drought planning exercises in various locations in the USA over the past several years. The usual outcome of such exercises is a greater awareness of the non-structural region-wide options that can be taken to reduce adverse drought impacts, often involving interagency cooperation, during periods of droughts.

Droughts affect industrial and business sectors in various ways, depending upon the nature of the particular industries. Production may decrease as a result of limitations on water availability. Water supply costs may increase. Hydropower production may be curtailed just when energy demands are increasing. Waterborne transport may be restricted. All of these events can lead to reduced economic productivity and even employment.

Drought mitigation measures include long-term programs and projects and emergency short-term actions managing water supplies and demands. A common long-term strategy for dealing with droughts is to store water in surface and groundwater reservoirs in times of high rainfall for use during periods of low flows. Reservoirs with large surface areas such as those on the Colorado and Missouri Rivers in the USA are designed for extended droughts, with storage capacities equivalent to several years of mean streamflow. Many other reservoirs operate in an annual refilling cycle, with sufficient capacity to store water during a wet season to meet demands during the dry season of the same year, thus providing little protection against an extended drought.

In addition to local or regional water supply storage, interregional water transport systems, involving aqueducts, canals and pipelines facilitate the importation of water from distant sources. Extensive regional water transport networks have been constructed throughout the world. However, the existing water in these sources usually cannot be increased during droughts. Thus, the operating policies of these reservoirs must include the definition of triggering indicators that identify the beginning of a drought, and release rules during droughts.

Long-term demand management or conservation programs promote the efficient use of water continuously during normal as well as drought conditions. Demand management strategies can include:

- the minimization of water wasted by seepage or leaks in conveyance and distribution systems,
- minimization of evaporation,
- efficient irrigation systems,
- drought tolerant crops,
- water efficient landscaping residential plumbing and industrial processes,
- re-use of treated wastewater,
- pricing structures designed to encourage efficient use,
- regulatory requirements for efficient water use practices, and
- public education programs.

Emergency supply augmentation during drought may involve:

- interregional transport of water by pipelines, trucks or other means,
- reallocation of water between users or purposes such as curtailing of irrigation use or hydroelectric energy generation to provide additional municipal water supply,
- drilling of wells, and
- desalination of sea water.

Although a variety of means are available for increasing supplies during drought, they are rarely well implemented (Dziegielewski et al., 1991). Most measures require fairly sophisticated preplanning. Many require forecasts (that are not available) of impending drought to trigger actions. Some require costly construction of facilities that are used only on an intermittent basis. Some require cooperation between independent operating agencies and public officials. Unfortunately, implementing agencies often do not understand pertinent concepts of risk management.

Ground water is often used to increase the supply of water during droughts. The search for additional sources can usually be carried out in a few months and allows the users to exploit these sources. There are, however, large areas where groundwater resources are either nonexistent or too uneconomical to use. The coupling of emergency wells, where they can be established, with the general distribution system gives the opportunity for tradeoffs among the users. The danger with emergency relief wells, however, is that after a drought they may continue to be used, and effectively increase the demand. If this happens, the systems will likely be more vulnerable than ever when the next drought occurs.

Emergency demand management is the most common and important solution. Demand management measures during drought include:

- public information and education campaigns,
- restrictions on nonessential water use,
- rationing programs, and
- emergency water pricing.

Public information campaigns are a particularly popular demand management strategy. Such campaigns: (1) encourage consumers to conserve water, and (2) provide them with information on how to do so. Research and past experience have shown that significant reductions in water use are typically feasible. But potential reductions vary greatly between regions and economic sectors. In general, the smaller the gap between available supply and demand and the larger the level of efficiency of a system the more vulnerable it is to droughts since by definition there are little or no means to adapt to reduced supplies. In many arid regions water use is often more efficient and hence additional reductions can result in greater hardships.

Environmental and ecological drought impacts

Environmental and ecological impacts of droughts can include:

- damage to ecosystems including plant, animal and fish species,
- loss of wildlife habitat,
- wind and water erosion,
- water quality effects; air quality effects due to dust and pollutants, and
- aesthetic and landscape quality changes.

Drought is a natural phenomena, and drought conditions have beneficial as well as adverse impacts. For example, dry periods reduce the vegetation that infringe upon wetlands. This reduces the tendency for vegetation to 'choke' the wetlands and leads to improved waterfowl habitat. However, effects on fish and wildlife population are often not evident until those born during the drought reach maturity. Unfortunately, impacts to ecosystems may potentially be magnified as a result of human response to drought.

Droughts affect water quality in various detrimental ways. These include increases in salt concentrations as normal flushing is disrupted by low streamflows. Bacterial counts can increase as water temperatures rise. Increased blue-green algae blooms mark significant changes in water conditions, including the pH and dissolved oxygen levels of streams, lakes and reservoirs. In coastal areas, decreases in freshwater flows can result in salt water intrusion into rivers, estuaries and aquifers.

Regional drought policies for increased sustainability

Developing a national or regional drought policy and plan is essential for reducing societal vulnerability and hence increasing its sustainability. Such plans should be based on the concepts of risk management. But drought planning is certainly complicated by the multitude of governmental and private entities involved, and the magnitude and diversity of drought impacts.

Wilhite (1993) outlines a framework for developing a national drought plan that consists of the following ten steps:

Step 1 – Appointment of a national drought commission

Step 2 – Statement of drought policy and plan objectives

Step 3 – Avoiding and resolving conflicts between environmental and economic sectors

Step 4 – Inventory of natural, biological and human resources and financial and legal constraints

Step 5 – Development of a drought plan

Step 6 – Identification of research needs and institutional gaps

Step 7 – Synthesis of scientific and policy issues

Step 8 – Implementation of the drought plan

Step 9 – Development of multilevel educational and training programs

Step 10 – Development of drought plan evaluation procedures

In the USA a regional or local drought preparedness study methodology was developed by the US Army Corps of Engineers and applied to several case studies in conjunction with the National Study of Water Management During Drought (Institute for Water Resources, 1994). The goal of this national study was to reduce drought impacts by collecting and improving on what had been learned about managing water resources for drought, and making this information available to water managers (Institute for Water Resources, 1991). Authorization of the study by the US Congress was motivated by severe droughts throughout the country in 1986-88 that continued in many places during the study.

Developing 'shared vision' decision support systems (see Chapter 7) and conducting simulation exercises for drought management involving all stakeholders and water utility agencies is an important feature of this methodology. The following sections highlight some recent drought experiences and mitigation measures taken in selected regions of the world. Readers can judge to what extent these measures may be sustainable.

Some experiences in drought damage risk mitigation in the USA

All regions of the United States periodically experience drought. The 'Dust Bowl' drought of 1933–35 and the 1950–57 drought in the Southwest, the 1964–67 drought in the Northeast and the 1976–77 drought in California are among the more significant droughts in terms of duration, magnitude and areal extent. Severe drought conditions covered large areas throughout the country during the late 1980s, particularly in 1988. The following discussion highlights drought experiences and mitigation measures taken in selected regions of the country during various periods between 1985 and 1992, particularly in the Southeast during 1985–86 and in California during 1987–92.

The 1985–86 drought in the Southeast USA

The Southeast USA experienced an agriculturally, meteorologically and hydrologically severe drought in 1985–86, just a few years after a generally less severe major drought in 1981. The drought began in 1985 in much of Alabama, Georgia, Tennessee, North Carolina and South Carolina (Golden & Lins, 1990). During 1985 the annual precipitation ranged from 65 to 95 percent of normal and the streamflow ranged from 50 to 80 percent of normal. Rainfall continued to be very low in the winter and spring of 1986. Record low streamflows were measured in parts of Alabama, Tennessee, Georgia and the Carolinas. During the first six months of 1986 a new low precipitation was set for Atlanta, Georgia. It was the second lowest January–June precipitation on record for Raleigh, North Carolina and Nashville, Tennessee. Cumulative streamflows through August 1986 were less than 40 percent of normal from Mississippi to North Carolina. Early 1987 continued to be dry, with high precipitation in the winter of 1987 generally considered to have ended the drought.

The 1985–86 drought in the Southeast severely affected agriculture, with estimated losses exceeding $1 billion (Golden & Lins, 1990). Many towns experienced water supply problems. Water quality problems occurred in streams and reservoirs throughout the region. Fish kills, odor problems and excessive weed growth resulted from low stream-

flow and high temperatures. Increases in salt concentrations were also a problem.

Most southeastern states took management actions during the drought and established a variety of contingency plans. In some places, particularly in Georgia and the eastern Carolinas, water use was restricted. Voluntary and mandatory water use restrictions were also implemented in many communities in Alabama, South Carolina, North Carolina and Tennessee. The Tennessee Valley Authority's hydroelectric power production was reduced by 50 percent during January–May 1986 to conserve water. The southeastern Power Administration, Alabama Power Company and other utilities also reduced hydroelectric power production, shifting to a greater reliance on their thermal-electric power plants.

The states of Georgia, Florida, Alabama and the US Army Corps of Engineers formed a drought management committee for the Apalachicola–Chattahoochee–Flint River basin. The committee coordinated many actions taken by the member organizations in response to the drought including reduction in reservoir releases for hydropower and navigation, implementation of water conservation measures and water use restrictions when necessary. Other states also created task forces to coordinate drought response efforts.

The 1988 drought in Western USA

In 1988 a drought spread through the West, Midwest and Northwest regions of the USA and portions of Canada. In June, 1988, the US Secretary of Agriculture declared 1,000 counties in 21 states to be disaster areas and thus eligible for federal assistance (Vlachos, 1990). Plans for diverting water from the Great Lakes into the Mississippi River were discussed. Forest fires occurred and agricultural losses were severe throughout the Great Plains states and Canada.

State-level drought plans and actions implemented in response to the drought of 1988 were generally short-term emergency measures intended to alleviate the crisis at hand (Dziegielewsksi et al., 1991). For example, Kansas established a hay and forage 'hotline'. Minnesota established a drought task force to address various problems. The Washington state legislature mandated the development of a drought contingency plan by the Department of Ecology for incorporation into the State comprehensive Emergency Management Plan. The US Army Corps of Engineers initiated extensive studies to update operating procedures for the reservoirs on the Missouri River.

The 1987–92 drought in California (USA)

The six-year 1987–92 drought in California is the worst six-year period on record (Dziegielewski, Garbharran & Langowski., 1993). During the drought, annual precipitation varied from 61 percent of average in 1987 to 86 percent of average in 1989. Annual runoff varied from 43 to 72 percent of average.

In the Central Valley of California, one of the most productive agricultural regions of the nation, agricultural water use declined only seven percent in 1991, with a loss in gross farm revenues of less than $600 million (Rich, 1993). The main reason for this relatively mild impact was an increase in groundwater use. Ground water generally contributes about 40 percent of the total California water supply, but accounted for 60 percent during the drought. About 1,300 new agricultural relief wells were drilled. Significant declines in groundwater levels resulted from the increased pumpage. Water from reservoir storage was also available during the early years of the drought.

Municipalities throughout the state implemented voluntary and mandatory water conservation programs. Plumbing retrofit programs and projects for use of reclaimed wastewater were also implemented. Hydroelectric power produced by California utilities declined from a normal level of over 30 percent of their total electrical generation to 18 percent during the first four years of the drought. Perhaps the most severe impacts of the drought were on fish (particularly salmon), wildlife and ecosystems that depend on the flow of rivers for their survival. The environment was impacted from the very beginning of the drought in 1987.

The drought also adversely affected forestry. The death rate of trees increased throughout the drought, with impacts expected to continue for decades. Forest fires also became a major problem.

In 1992 the Governor of California proposed a comprehensive new water policy to lessen the impacts of future droughts. The fundamental concepts reflected in the plan are:

- protection of environmental resources,
- constructing additional storage facilities,
- encouraging water marketing,
- promoting water conservation, and
- addressing the problem of groundwater overdraft.

The most significant institutional water management change to stem from the drought was the creation of the California Drought Water Bank (Rich, 1993). The state water bank is authorized to acquire water in three ways: (1) by paying farmers to not irrigate their fields, (2) by purchasing surplus water from local water districts, and (3) by paying farmers or water districts to use ground water instead of surface water.

Dziegielewski, Garbharran & Langowski (1993) investigated lessons learned from the California drought. They summarized them as follows:

- The nature of social, environmental and economic impacts of a sustained drought points to a need for careful and more realistic drought planning.
- Severe drought can change long-standing relationships and balances of power in the competition for water.
- Irrigation can provide complementary environmental benefits.
- Land-use regulation must be the mechanism for urban growth management policies which deal with limited water supplies.
- The success of drought response plans should be measured in terms of the minimization and equitable redistribution of actual impacts (as opposed to water shortages), but there is much to be learned about the best ways of accomplishing this.
- Severe droughts can expose inadequacies in the performance and roles of state and federal water agencies.
- The overall success of water rationing plans depends on their design and reliance on increases in water rates.
- Mass media can play a positive role in drought response.
- Market forces are an effective way of reallocating restricted water supplies.

Experiences in drought damage risk mitigation in the Segura River basin (Spain)

The Segura River basin covers a large portion of the southeastern portion of Spain and is the most arid region in Europe, with a mean annual precipitation on the order of 300 mm. On a characteristically humid continent with a significant degree of development, a region like the Segura River basin with a warm climate and abundant sunshine has an important economic advantage: complementarity. Tourism and semi-tropical cash crops are the basis for the economic development of the area.

The southeast region of Spain, however, has extremely limited water resources, with 90,000 m^3/year covering a region of 22,000 km^2 and a population of approximately 3 million. Irrigation has been practiced since Roman times with improvements during Arab domination. The irrigated area by the end of the 18th century was 58,000 hectares. High water table levels in the flood plain of the Segura River basin, where the oldest irrigation areas are located, forced the users to rely on constructed drainage systems. These systems were developed such that return flows either discharge to the river or are re-used for irrigation without any dilution. Despite the size of the irrigation area, consumptive use was only 30 percent of the total water resources.

The system worked without major problems until the beginning of the 20th century. Then the availability of water increased, mostly due to the construction of large dams and intensive use of the aquifers hydraulically connected with the river. The result was a doubling of the irrigation area. The drainage system allowed the water to be re-used five consecutive times with consumption around 70 percent.

During Spain's period of economic development (around 1970) a large water aqueduct was built to transport the water from the Tajo River in the center of Spain to the southeast. The Tajo-Segura Aqueduct, 350 km long, was designed with a capacity to provide an additional amount of 100,000 m^3/year to the Segura basin, although only 60,000 m^3/year could be legally used. The main goal of this project was to add 100,000 hectares of irrigated land to the river system. These new areas were to be highly technical and modern land parcels for the development of citrus, vegetables and other cash-crops, and to ensure that a water supply existed for the permanent population and tourism. When the trans-basin diversion works were finished in 1973, irrigation was incorporated in new areas and in many cases used ground water in anticipation of the imported water.

At present the Segura River basin has 197,000 hectares under irrigation, but the promised water from the diversion has never arrived due to socio-political reasons. At a maximum (in a wet year) the amount of water diverted from the Tajo reached 32,500 m^3, but the annual average has not reached 20,000 m^3 in 12 years of operation. This deficit from the expected 60,000 m^3/year has produced a lack of equilibrium that clearly affects the sustainability of the total system. The annual consumption demands now reach 156,300 m^3 of which 136,800 m^3 correspond to irrigation and 19,500 m^3 to urban water demand. When the supply (including the diversion from the Tajo) does not reach 115,000 m^3 the difference is made up, in the short term, with an over-exploitation of the aquifers.

This groundwater over-exploitation has reduced the Segura baseflow to almost zero. The extensive re-use of river water has caused water quality to reach unacceptable levels in the irrigated area in the Segura, with salt concentrations above 8,000 ppm. This has produced an accelerated process of salination and the consequent abandonment of irrigated parcels. This is happening mainly in the oldest irrigation sites, which paradoxically are those that have the most senior water rights. Those farmers have small farms that average 0.3 hectares, the owners living on the plots, whereas the new developments are large plots consisting of tens or even hundreds of hectares. The social problem is significant because it is impacting the poorest sectors of the population.

The alleged main reasons for denying the transfer of water from the Tajo are a drought and climatic change. However, an analysis of the time-series of inflows to the upper Tajo reservoirs does not support these claims. The main reasons that the storage in the Tajo reservoirs remain at low levels are the releases for hydropower production and the use of water for dilution of wastewaters from Madrid. These two uses require reservoir releases in wet years that otherwise could be stored. Perhaps the main underlying reason is that it is not socially feasible to divert water due to the opposition from the Tajo region (Castilla-La Mancha). This is a large sparsely populated and poor region with a cold climate that has lost about half of its population in the 20th century through emigration.

To overcome the opposition to the diversion, economic compensation (in the form of new irrigated areas) in the Tajo region was proposed. However, this approach did not consider the economic situation from national and global points of view. The main crops that could be grown in the Tajo region (for climatic reasons) are subject to international market prices that are well below local production costs. The development of new irrigation sites in the Tajo to produce such crops would result in a heavily subsidized agriculture which is strongly opposed by the EEC and in the future by GATT agreements.

There are several lessons for developing countries to learn from this yet unsolved problem. The Tajo-Segura project was conceived and built when Spain was developing and under an authoritarian government. The decision to build the diversion was mainly based on technical merits at the time. These technical considerations are still valid but do not reflect all the economic, environmental and social factors that should be taken into account in a water resources project. There was no planning for exploitation, no consideration of the environmental impact in both basins, and no full study of the legal implications of a trans-basin diversion. It was built under the assumption that the diversions would be made 'when surpluses existed in the Tajo' without defining what those surpluses were to be. The logical reaction in the Tajo region was to declare itself in 'perpetual drought'.

The fundamental mistake, and a typical authoritarian governmental error, was to neglect the socio-economic considerations of both regions. To carry out a diversion even under a democratic regime is extremely difficult and complex, not for technical reasons but for socio-economic and legal considerations that can make the diversion politically infeasible. In a drought situation, the reallocation of water resources involves mainly socio-economic, legal and management problems rather than hydrologic or engineering ones.

The Tajo-Segura diversion today remains like a dry canal in the Castilla plateau. Its planners and builders have generated an unsolvable environmental and desertification problem. This is a good example of unsustainable development. The water resources of the region were exploited above their maximum level, thereby creating irrigated areas that have subsequently become sterile and saline.

Experiences in drought damage risk mitigation in Peru

Peru's 23 million people inhabit an area of 129 million hectares, greater than the combined areas of France, Italy and Germany. About 10 percent of Peru's area is located along a narrow coastal strip between the Pacific Ocean and the foothills of the Andes. This area is largely arid with an annual average rainfall of only 38 mm. However, it enjoys moderate temperatures and fertile soils and receives water from 53 rivers that flow from the Andean mountains. It also has limited amounts of ground water in some areas.

The availability of surface water in Peru is equivalent to a mean annual flow of 64,200 m^3/s, or approximately 5 percent of the world's runoff. But the bulk of these resources (some 98 percent) is located in the Amazon basin. The coastal region is entirely dependent on uncertain and highly variable seasonal supplies, with over three-quarters of the water flow occurring between January and April.

The scarcity of water in the fertile coastal region has led to heavy investments in hydraulic infrastructures to regulate the water from Andean basins and to convey and distribute it to coastal users. Some of these systems are hundreds of years old. Currently, some 717,000 hectares are irrigated in 53 coastal valleys. Of these, only about 320,000 hectares have guaranteed water for irrigation purposes.

By 1992, following several years of virtually no public spending for maintenance or rehabilitation of water-regulating infrastructure, most of the systems faced the risk of sudden failure. Water delivery became more irregular, water quality deteriorated and water conflicts grew. Even in areas where water was scarce, it continued to be used wastefully.

Although Peru's existing water law specified priorities for water use, mandated that water tariffs cover both operating, maintenance and repair (OMR) and construction costs, and

established a sophisticated institutional framework, the system of priorities for water use has proven unworkable. Water tariffs typically do not even cover OMR costs and the public institutional structure is weak. Despite low water tariffs, many users were unwilling to pay them and the government knew that it would be extremely difficult to increase water charges substantially or to create strong institutions to plan, assign, monitor and enforce efficient water use. Moreover, it was unwilling to devote substantial public resources to subsidize the hydraulic infrastructure.

To address these problems, the government of Peru has proposed, as the centerpiece of its water management development strategy, a new water law modeled along the 1981 Chilean water code, which in that country has successfully improved water delivery and use, stimulated private investment and reduced water conflicts. Under the proposed Peruvian law, existing water users are to be given property rights to water use without charge. Rights to new or unallocated surface water are to be distributed via a public auction. The rights may be traded at freely negotiated prices provided that the trade does not reduce water availability to others and that there is enough water to maintain a minimum ecological flow and the accustomed quality of life in cities and towns. Rights may also be mortgaged or leased. The law would prohibit alterations to the detriment of flora or fauna; however, rather than proposing specific sanctions and fines, it defers to the environmental code to set and enforce water quality standards.

CONCLUSIONS

The natural variability of water supplies, including its extremes manifested by floods and droughts are natural hazards and, as such, are not completely avoidable; however, planning, forecasting and early warning can significantly reduce human and economic losses. Water resources planning, beginning at the basin level, should emphasize and implement cohesive efforts by national and city governments and local entities to minimize the fragmentation of responsibilities in combating and preventing the destructive effects of these events. Most of the adverse impacts of droughts can be substantially reduced with proper management. However, the majority of water supply agencies do not rely on aggressive drought management programs, with *ad hoc* response measures typically being implemented when droughts occur.

Sustainable development, with regard to water resources extremes, should rely on the participation of all users in the decision-making process, financial accountability and cost-sharing and balanced structural and non-structural measures, with greater emphasis on non-structural measures. Proactive, rather than reactive, measures are the key to the sustainable development of water resources.

THINGS TO REMEMBER

- Anyone trying to predict what may happen or be observed in the future, especially the distant future, should know that such predictions are at best risky, possibly uncertain, and subject to all sorts of surprises that cannot even be imagined.

- Sustainability implies a condition in which the threats to society, as well as the threats to water, are managed in such a way that people are prepared and able to cope with them when they occur, and that their frequency and severity are decreasing over time.

- The prevention, management and control of disasters have a high proprity in the achievement of sustainability. However, it is usually neither politically feasible nor economically possible to eliminate all potential hazards or to design all water resources systems to withstand any coceivable extreme event.

- Risk management has been formalized in recent years. The first step in preparing for potential disasters is to identify the potential hazards existing in a region, and evaluate their probabilities of occurrence.

- In risk assessment the analyst often attempts to answer the following: What could go wrong? What is the likelihood that it will go wrong? What would be the consequences?

- The analyst then builds on the assessment process by seeking answers to the second set of questions: What can be done, i.e., what options are available? What are the associated tradeoffs in terms of all costs, benefits and risks? What are the impacts of current management decisions on future options?

- A total management approach must include risk management as part of the overall system management of possible system failures. Of equal importance is the total involvement in risk assessment and management by everyone concerned.

- Strengthening of land-use and urban planning in many regions would prevent, with a minimal use of structural measures, situations that will require construction of hydraulic flood control structures after an area has been developed. Small easements today may substitute for expensive hydraulic structures in the future.

- The effect of drought on public water supplies necessitates cooperation between water users and local, regional and national public officials. But, since droughts are infrequent in many areas, water managers are faced with dealing with situations during droughts for which they typically have little or no past experience.

- Long-term demand management or conservation programs can promote the efficient use of water continuously during normal as well as drought conditions.

- Developing a national or regional drought policy and plan is essential for reducing societal vulnerability and hence increasing its sustainability. Such plans should be based on the concepts of risk management.

11 Equity, education and technology transfer

A major aspect a of sustainable water resources planning and management effort is that those people who will be affected should participate and equitably share in the policy formulation as well as in its benefits and costs. To accomplish this requires an educated and informed citizenry. The results of the process should be the development of the technology needed to do the job – a technology that is appropriate for the region and is capable of being effectively used and maintained.

In the early stages of many major water resource systems projects some individuals will inevitably be inconvenienced by construction and resettlement activities. Resettlement can involve involuntary moves from ancestral homes and traditional living conditions to areas that are unfamiliar – a cost difficult to measure and rarely fully compensated for by those benefiting from the increased irrigation, hydropower, flood control, and other project purposes. Changing streamflow regimes caused by dams or major diversions to offstream water users can also force major adjustments in the way individuals work the lands downstream from those projects.

The development of a water resource system infrastructure has often been of greater benefit to far away users than to the local people. Even worse are the cases in which projects have not yielded the desired effects, or in which projects function only for limited times. An example is a dam built to supply water for irrigation in an arid country where excessive siltation may cause the reservoir to rapidly fill with sediment. What happens if the reservoir becomes so full that the system can no longer operate? A better choice, no doubt, may have been to improve existing methods of agriculture rather than to introduce, through a system that depends on the dam, an agriculture that can last for only a limited time. A silted up reservoir not only becomes an essentially useless capital investment, it also can disrupt the society for whom it was designed to serve. During the lifetime of a dam local farmers will adjust to irrigation depending upon the reservoir, changing their life styles and working habits accordingly. While the dam is still in operation, the traditional approaches can be forgotten and experiences in sustainable and adapted agricultural practices can be lost. When the reservoir becomes filled with sediments, the old approaches will be needed, but may be long forgotten. Such disruptions in life styles and in supply can obviously affect the stability of whole regions.

To be sustainable, a project must perform reliably and the transition to new technologies and management practices must proceed in an orderly and equitable manner. Continuity and confidence in the new systems are prerequisites for sustainability, as are a proper respect for operation rules and for maintenance of the physical infrastructure.

EQUITY AND PEOPLE INVOLVEMENT

The tasks of political decision-makers include the weighing of benefits against any negative consequences and to set priorities. They have a special responsibility for water resources projects. A project planning process based on sustainability concepts must be a part of an overall development plan, one that embeds the water resources planning process in the general development strategy of a region or country. However, such a general plan should not be developed by experts and administrators alone. Rather, the public should be involved as much as is feasible, so that decisions are supported by consensus.

To quote from the address of Charles, Prince of Wales, to the World commission on Environment and Development in London, in April 1992:

> Starting with people, analyzing their needs, taking account of their culture and traditional practices, making certain that the roles of all sectors of the community are understood, and above all, to ask people to frame their own, local, environmental goals are all prerequisites to satisfy solutions (of development and environmental problems).

Involving every person who has an interest in a fair and orderly decision-making procedure should lead to mutual problem solving instead of tug-of-war among different interest-groups. Most developed regions have the legal framework for conflict resolution among different water users through a continuous process of problem recognition, political legislation and legal and administrative actions. In both developed and developing regions, non-governmental organizations concerned with environmental and social issues can play an important part in helping to provide such frameworks. They speak for at least some of the people, and they can help provide a balance between governments and the citizens affected by proposed or even existing development projects.

In many of the poorest regions of the world, water ranks low in national priorities. Hence, they do not have balanced water management standards comparable to those with greater resources. Traditional ways of helping such regions, by providing assistance for water resources development projects similar to those that exist in other parts of the world, may not be appropriate. More important is the need to develop techniques based on local skills and technologies. Engineers and managers alike have the challenge to look for local ways in which the development goals can be reached in an efficient manner that is also adapted to local conditions. Standards or techniques from other countries should not be imposed arbitrarily. Instead, local engineers and technicians should learn to identify the requirements in these areas from consultations with the local people, and to find ingenious and adapted ways to meet them. To quote from the statement made by Mrs Traore (1992) during the Dublin Water conference:

> Despite the economic and financial difficulties confronting them, the Third-World countries, particularly rural communities and women, can play a decisive role in protecting the environment and in managing the natural resources, including water. The questioning currently taking place at international levels and within individual countries is creating the necessary conditions for a new

partnership with populations who need training and information on the mechanisms of development, including funding. In the meantime, it is premature to ask them suddenly to cover their own costs in all fields, at the risk of aggravating poverty and causing further severe social and political conflict

> Humanity has the means of avoiding such a situation but it will only be able to do so if the development theoreticians drop some of their set ideas, particularly as regards human development, and give the communities, particularly women and young people, the freedom to express their own points of view and take their own decisions.

There are encouraging signs that people in many parts of the world are approaching their water supply problems in an adaptive way, and that these are based on the wishes and the financial and technical capabilities of the people themselves. Dedicated planners and engineers from developed countries are learning how to help people better understand how to solve their own water resource management problems and how to achieve increased benefits from that resource. Outside assistance can affect attitudes towards the use of multiple smaller more sustainable projects as opposed to much larger and more expensive engineering works, which unless well understood and accepted may not be used effectively by the people for whom they were presumably constructed.

Involvement of people is important for the impact assessment of any structures that can involve changes in local societies. Design and construction of hydraulic structures used for flow regulation and storage used to be the exclusive prerogative of engineers. Increasingly, engineering works are affected by non-engineering considerations because of their potential impacts on environmental, social, cultural and aesthetic values. Whether or not a project is beneficial is often a question of 'beneficial to whom?' For example, planning of a hydro-project should not only involve the beneficiaries of the project, but should also require informing and involving all those directly or indirectly affected. Mechanisms that look after the interest of those affected have to be incorporated into any development and management plan that is to provide increased sustainability.

INTRATEMPORAL AND INTERTEMPORAL FAIRNESS

Recent work of Matheson (1997) addresses intratemporal and intertemporal fairness with respect to sustainability criteria. He identifies overall fairness as a weighted combination of proportionality, equality, and need based measures and offers this as one of many criteria for selecting sustainable development projects or operating policies. This section defines these fairness measures.

The literature on sustainability and equity refers to *intra*-generational fairness and *inter*-generational fairness. *Intra*-generational fairness is fairness within a generation while, *inter*-generational fairness is fairness between generations. Intra-generational and inter-generational fairness comparisons are simply intratemporal and intertemporal fairness comparisons over at least one generation. Intratemporal comparisons can be made across all groups during any given time step. Intertemporal comparisons can occur, for a given group, across all time steps for which that group exists.

Consider a situation in which there are a total of G groups and each group g is affected by various impacts i. For this discussion we will consider only one type of impact (either positive or negative – a benefit or a cost), realizing we can apply our measures to each such impact. Define $A(i,g)$ as the magnitude of the actual (predicted) impact i to group g that will result from some action. A is the level of impact that the party (or group) actually will get. Define $E(i,g)$ as the magnitude of the expected (beneficial) impact i group g needs or deserves, based on, for example, their proportionate size as a group, or what they sacrificed for the project, or some other attribute. Let $Z(i,g)$ be the magnitude of the impact i group g requires to satisfy its needs or wants. Measures identified by Matheson (1997) that involve equity, equality, and need components of fairness, $F(i)$, are given in equations (11.1–11.3):

(a) Sum of absolute deviations of impact i from an equity allocation;

$$F_1(i) = \sum_{g=1}^{G} |E(i, g) - A(i, g)| \tag{11.1}$$

(b) Sum of absolute relative deviations of impact i from an equality allocation among all groups (denoted by the indices g and h);

$$F_2(i) = \sum_{g=1}^{G} \sum_{h=1}^{G} \left[\frac{|A(i, g) - A(i, h)|}{2G^2 \bar{A}} \right] \tag{11.2}$$

where $\bar{A}$ is the mean impact i over all groups.

(c) Sum of absolute undesired deviations from an impact i allocation that satisfies the desires of all groups:

$$F_3(i) = \sum_{g=1}^{G} \left| [Z(i, g) - A(i, g)]^{\text{undesired}} \right| \tag{11.3}$$

If the impact is positive (e.g., benefits) then the values of the expressions included in the sum of Equation 11.3 are only the positive ones, otherwise they are the negative ones.

If A is the actual impact experienced by a group, and we want to make A as close to 'fair' as possible, we would compare A with a reference distribution. The choice of reference distribution depends on what one defines as fair, or as a fair allocation objective. Using the three equations above, Matheson identified equity (or proportionality), equality and satisfaction of need measures of fairness. The deserved impact, E, is the reference distribution if equity were the fairness objective that is to be maximized. The perceived need, Z, is the reference distribution if satisfaction of need were the fairness objective that is to be maximized. All As are compared with Es or Zs to see how close the actual allocation of impacts are either equitable or satisfy perceived needs. The As are compared with each other to measure the extent the actual allocation of impacts are equal.

Equation 11.1 represents the magnitude of fairness resulting from impacts which deviate from an allocation that is proportional to each group's contribution. Note that whether a desired or undesired impact A is greater than or less than E, an equitable allocation objective attempts to make A, as close to the expected, or deserved impact, E, as possible. Overshooting is given the same weight or importance as undershooting, hence the absolute value of deviations are used. These could be altered if desired.

To have perfect equality, it would be required that everyone's (or all groups') impacts As would be equal. We evaluate the equality objective by comparing all groups' impacts. Equation 11.2, known as the Gini Index or Gini Coefficient, is commonly used to measure deviations from perfect equality. The two group indices, g and h, are required for Equation 11.2 as the numerator of this equation is the sum of all possible pairwise comparisons of impact magnitudes between groups. The Gini Index's magnitude represents the decrease in fairness resulting from impacts that deviate from being allocated equally among all G groups.

Equation 11.3 is presented as one way to measure the fairness in an allocation of impacts that deviate from meeting each group's perceived need for that impact. It should be noted that other methods of operationalizing a need fair allocation objective, such as a binary expression of whether or not a need is met, may be possible.

Equations 11.1–11.3 have values that may increase in magnitude from zero, where zero corresponds to complete fairness as defined by each fair allocation objective.

If these equations are to be used in the context of sustainable project selection, the equations must be expanded to address four major concerns. First, these measures need to be formulated in a way that accounts for the dimensions required for both intratemporal and intertemporal comparisons. Second, projects are likely to distribute multiple impacts among groups, and a measure of fairness should account for this in some way. Third, Equations 11.1–11.3 represent three different aspects of fairness evaluations, and a single overall fairness measure should incorporate all of these objectives. Not all people evaluate fairness by equity, equality or need satisfaction alone. Any overall measure should reflect the variability in the emphasis placed on the different fair allocation objectives by the groups who are affected by a project.

Consider a situation in which there are P project alternatives, each distributing I different impacts to G groups over T time steps. Thus, each project alternative may be thought of as having $G \times I \times T$ impacts. The magnitude of impact i accruing to group g, during time step t, resulting from project alternative p may be written as $A(i, g, t, p)$. For each group, g, the impact i it deserves, may vary with time step t and is written as $E(i, g, t)$. Additionally, a group's need for a particular impact may also vary with time and is written as $Z(i, g, t)$. Equations 11.1–11.3 may be rewritten to correspond to this generalized problem. Equations 11.4–11.6 represent the three fairness measures expanded for the intratemporal case with any number of impacts. Equations 11.7–11.9 represent the three fairness measures expanded for the intertemporal case, with any number of time steps.

(a) Average intratemporal fairness measure of project p which is the weighted sum of absolute deviations from an equity impact allocation for all groups, g.

$$F_1(p) = \frac{1}{IT} \sum_{t=1}^{T} \sum_{i=1}^{I} \left[w_i \sum_{g=1}^{G} \left| A(i, g, t, p) - E(i, g, t) \right| \right]$$
(11.4)

(b) Average intratemporal fairness measure of project p which is the weighted sum of absolute relative deviations from an equal impact allocation for all groups:

$$F_2(p) = \frac{1}{IT} \sum_{t=1}^{T} \sum_{i=1}^{I} \left[\frac{w_i \sum_{g=1}^{G} \sum_{h=1}^{G} \left| A(i, g, t, x) - A(i, h, t, p) \right|}{2G^2 \overline{A_{itp}}} \right]$$
(11.5)

(c) Average intratemporal fairness measure of project p which is the weighted sum of absolute undesired deviations from an allocation that meets the needs of all groups

$$F_3(p) = \frac{1}{IT} \sum_{t=1}^{T} \sum_{i=1}^{I} \left[w_i \sum_{g=1}^{G} \left| [Z(i, g, t) - A(i, g, t, p)]^{\text{undesired}} \right| \right]$$
(11.6)

(d) Average intertemporal fairness measure of project p which is the weighted sum of absolute deviations from an equitable impact allocation for all groups:

$$F_1'(p) = \frac{1}{GI} \sum_{i=1}^{I} \left[w_i \sum_{g=1}^{G} \sum_{t=1}^{T} \left| A(i, g, t, x) - E(i, g, t) \right| \right]$$
(11.7)

(e) Average intertemporal fairness measure of project p which is the weighted sum of absolute relative deviations from an equal impact allocation for all groups:

$$F_2'(x) = \frac{1}{GI} \sum_{i=1}^{I} \left\{ w_i \sum_{g=1}^{G} \left[\frac{\sum_{s=1}^{T} \sum_{t=1}^{T} \left| (i, g, s, p) - E(i, g, t) \right|}{2T^2 \overline{A_{igp}}} \right] \right\}$$
(11.8)

(f) Average intertemporal fairness measure of project p which is the weighted sum of absolute undesired deviations from an allocation that meets the needs of all groups:

$$F_3'(p) = \frac{1}{GI} \sum_{i=1}^{I} \left[w_i \sum_{g=1}^{G} \sum_{t=1}^{T} \left| [Z(i, g, t) - A(i, g, t, p)]^{\text{undesired}} \right| \right]$$
(11.9)

In the above equations the weights, w_i, are assumed constant for each impact i; they may not be. The variables $\overline{A_{itp}}$ are the average predicted impact values for all groups given a particular combination of impact i, time step t, and project p,

$$\overline{A_{itp}} = \frac{1}{G} \sum_{g=1}^{G} A(i, g, t, p)$$
(11.10)

and $\overline{A_{igp}}$ are the average predicted impact values for all time steps given a particular combination of group g, impact i, and project p,

$$\overline{A_{igp}} = \frac{1}{T} \sum_{t=1}^{T} A(i, g, t, p)$$
(11.11)

Equations 11.4–11.6 are based on the three fairness allocation objectives applied to the impacts distributed among groups during a given time step which are then averaged over all time steps. Equations 11.7–11.9 are based on the three fairness allocation objectives applied to the impacts distributed over all time steps for a given group which are then averaged over all groups. These fairness measures may be further combined into overall measures of intratemporal and intertemporal fairness by a weighted average approach.

These fairness measures merely illustrate an approach for quantifying, to some extent, equity. Other measures may be more appropriate in a particular situation. Any fairness measure that does not accurately describe the perceived fairness of the groups impacted will not be considered relevant, or fair. Uncertainty may arise due to prediction errors, e.g., errors in the predictions of the impact predictions over time. Impacts may increase, decrease, remain constant, or be some combination of these trajectories over time, and the number of impacts that affect each group may also change over time. The approaches discussed in Chapter 4 can be used to assess the sustainability of any project based on these fairness measures.

INTERNATIONAL EQUITY ISSUES

Although water-related problems are usually local or regional, where rivers or lakes cross international boundaries, or where seas separate countries, the problems of water consumption and pollution become international problems, and as such they require joint approaches toward finding acceptable and effective solutions. Sustainability criteria should not only guide conflict resolution within countries but should also apply to international water-related disputes.

Most of the major river systems of the world are shared by two or more nations. The interests of multiple countries sharing common waters of a river can be quite different. The lower country may desire clean and unlimited amounts of water for its own development purposes, whereas the upper country may want to use the water of the river to its own full benefit, including (commonly) the disposal and transport of pollutants. Important sources of conflicts are water supply, river pollution and contamination of ground water. Such conflicts must be resolved through treaties, such as the ones which regulate the use of the Nile waters between Egypt and Sudan and the pollutant level in the Rhine River at the border between Germany and The Netherlands.

International conflicts introduce considerable risk for the proper functioning of water resources systems. The water supply of a region or of a country can be the most vulnerable part of the region's or country's economy and possibly its defense system as well. As a result, agreements among multiple regions or countries can be especially sensitive issues for future generations to maintain or resolve. The Ruacana project between Namibia and Angola is a case in point, where sabotage and terrorist activities have destroyed much of a

burgeoning bi-national project. Another case has been the Cabora-Bassa scheme on the Zambezi River for Mozambique and South Africa. Although the major structures of that dam and powerhouse remained intact, some 1000 km of power transmission lines were for many years periodically sabotaged.

The growing body of national and international legislation on water, pollution and the environment presents an opportunity and a challenge for future water managers. Its effective use requires broad training and exchanges of professional information. A world-wide exchange of legislative experience with local, regional, national and international problems of water supply and pollution control would be useful. Various principles of sustainability might be translated into legal requirements, perhaps at the international level, that would at least, discourage, if not prevent, the implementation of non-sustainable decisions based on special local, regional or even national interests. Because integrated management of river basins and estuaries, inland seas and lakes, together with their ecosystems, should be promoted within local, national, regional and across international boundaries to share common resources, the establishment of a universal code of ethics and standard legal operating procedure might be worth considering

EDUCATION, TRAINING AND TECHNOLOGY TRANSFER

To improve the sustainability of our water resources and associated ecosystems, a greater commitment is required by all of us. But the capacity to plan and carry out needed actions is as much a prerequisite as is the political will to do it. This can only be achieved by an informed and actively involved public. The political support needed will often come as a result. Governments, professional and voluntary organizations and the media must debate and educate each other on the critical issues, especially about the links between environment, population and development. To transfer effectively information between professionals and the public, professionals, including engineers, must become more proactive, contributing their professional expertise to an informed discussion of the positive and negative aspects of any water resources development project.

Commitment alone, however, is not sufficient. Local communities to entire countries, as appropriate, must have both the personnel and the know-how to cope with the problems involved in their particular water resources development (Alaerts, Blair & Hartvelt, 1991). They must also have, or develop, the capabilities of technicians to operate and maintain sustainable systems at local levels. A necessary condition is that appropriate institutions must exist to carry out planning, operation and maintenance of water resources systems. To be sustainable, a system must satisfy local administrative requirements. Its managers must be trained to manage the system taking into account the customs and legal infrastructure of the region. For this, information is needed so that the right methods can be found.

Improving the water consciousness of every water user and the capabilities of those responsible for water management within any region can be a challenge requiring extensive educational programs instituted at all levels of society. Water consciousness at the grass-roots level fostered through all stages of education can help ensure sound handling of scarce water resources, especially in regions that are short of water. Public education is needed for the management and protection of ground water, or for rural water schemes created by regional authorities. Programs should be established promoting good health and domestic hygiene practices. Unless their importance is understood by the public, institutional controls will not work.

The need for people to be broadly educated on water issues is particularly urgent when the people themselves are part of the water distribution system, as is the case for domestic water supply and in irrigation. Only when the farmers, for example, understand the need for water conservation will they be interested in water management for efficient crop irrigation.

Each nation must inevitably come to depend primarily on its own professionals to provide the know-how and experience required for water resources development. However, it may originally depend to some extent on expert advice brought in from other countries. In order to develop the skills needed to effectively solve design and water management tasks, the technical abilities of its engineers must be upgraded continuously through education of the most qualified young people and by continuously improving the professional abilities of the practicing professionals. The teachers required for this purpose should, to the greatest extent possible, be local persons who have been trained on actual projects, and who have supplemented their knowledge by advanced studies in high quality schools.

Teachers in local schools should be able to instruct students not only in their specific fields, but they should also provide information about the impact of development on the environment – both factually and philosophically. One

expects of them an awareness and concern for the fragility of natural ecosystems that they can transmit to their students. A good basis for this is an expanded systems approach, one that incorporates environmental issues into a multi-objective decision problem and leads to an appreciation for multidisciplinary cooperation and integral management. Finally, academic programs should be structured to be effective in meeting the demands of society, and teachers should be trained who are capable of implementing them.

To keep informed of new developments, teachers must have access to current information sources. Two ways are (1) participation in regional and international meetings and (2) through other means of technology transfer, e.g., dissemination of written information. For the teaching of teachers, many countries with well-developed training programs offer these to developing countries, and local efforts are supported by material provided by the international associations. A good example of this, at the international level, is the UNESCO/IHP Humid Tropics Programme and its 'popularized' document series covering a broad range of subjects.

Teaching and research go together. Much of the research needed for solving problems associated with ecology, economics, environment, hydrology, climate change and regional water resources systems planning (such as for development projects of water supply and sewage disposal systems for mega-cities) requires extensive resources in equipment and expertise from many different fields. Team efforts and research cooperation are often required on a national, and even an international, scale. Research is also necessary on integrated water resource management methods, for both rural and urban areas, that are adapted to the particular demands of different regions of the world. To serve these purposes the establishment of coordination centers has been proposed by UNESCO, e.g., the UNESCO/IHP Humid Tropics Programme. In these centers priorities of research and development can be established for a region. These research priorities can then be addressed by teams consisting of many different specialities.

In regions or countries lacking adequately trained personnel, outside expertise may be of some help in enabling local professionals to solve local problems. The international community of scientists and engineers is challenged to give the needed help unselfishly, and where information is not available, to coordinate and support national and international research. Professional societies can often facilitate the exchange of information and the interaction within countries and among international scientists.

A ROLE FOR PROFESSIONAL SOCIETIES ——

A key to the sustainable development of water resources is the existence of sufficiently well-trained personnel in all of the fields that are involved in the development process. In regions or localities where such a capacity does not exist, it must be developed. Professional societies can help facilitate the training of these professionals. Where engineering and scientific organizations exist, and have strong backing by professionals, they can have a considerable influence on the development of a region's water resources.

In regions, in particular in the developing world, where suitable organizations bringing engineers and scientists together do not exist, professionals should work to create and participate in regional or national professional societies. The important role which associations such as the Institute of Civil Engineers in Australia and England or the American Society of Civil Engineers in the USA have played in the development of their countries provide examples. Well-functioning engineering societies can help in the development process by providing a whole range of services. These services can include:

- providing a mechanism for the continuous updating, through journals, seminars and training courses in all aspects of water development, of the professional knowledge and skills of their members, and thus they prevent professional isolation.

- producing manuals of practice for planning, operation and maintenance, as well as for detailed engineering design.

- acting as links between institutions, such as public water and sanitation agencies, irrigation districts and private manufactures, consultants and other companies active in the sector.

- collaborating with national decision-makers in defining national policies in their sector, setting realistic targets and standards both for engineering practice and for equipment and material, and codifying good practices in utilities, government departments, consultants, manufacturers, universities, etc.

- promoting national and international exchange and cooperation in the areas of research, training, technology and overall strategy by transferring experiences of experts from one locality, region or country to another.

International professional Non-Governmental Organizations (NGOs) as well as the relevant UN agencies have served as links between professionals in developed and developing countries. Numerous cooperative projects and activities testify to the success of this approach. For example, the International Association for Water Quality (IAWQ) has given moral and technical support to national professional societies in Thailand, Malaysia and South Korea. The International Water Supply Association (IWSA), in partnership with IAWQ and the International Solid Waste Association (ISWA), has formulated a project for improving professional societies in selected developing countries.

By a systematic program of organization and management the professional societies seek to improve the competence of their members. Their short courses and other programs of technology transfer in subjects such as water supply and treatment, sanitation and solid waste disposal, as well as more advanced topics in computer modeling, provided by the members of the associations, serve to help keep professionals aware of the current state of science and technology as well as current standards of engineering practice. Through international associations of professional societies, local members can be integrated into the broader international community of professionals.

A significant role can be played by the international associations working in water-related fields. The cumulative experience of all engineers and scientists participating in the activities of these associations forms the solid basis on which water resources development rests. The dissemination of this knowledge to colleagues around the world has been and continues to be one of the most important tasks of the international associations in the water field. Foremost in this process are scientific meetings of various kinds. In addition, some associations also provide other services, many of which are especially intended for their members in developing countries. Some have even made service to these members a prime responsibility.

Years ago the Permanent International Association of Navigation Congresses (PIANC) established a special committee for finding ways of helping scientists and engineers in developing countries. An indication of this growing concern is that other associations are following this lead. The International Association of Hydraulic Research, for example, has two committees that are directly involved with such questions: a Committee on Continuing Education and Training (which has explored different methods of knowledge transfer and made its findings available to all members), and a Consultative Panel (which has the function of finding ways and means of intensifying, for the benefit of its membership, interactions between the association, national professional societies, the UN Organizations, and the large funding agencies such as the World Bank. The International Association of Hydrogeologists has its 'Burdon Commission' on developing countries with comparable objectives. The International Water Supply Association has followed the recommendation of its Committee on Cooperation and Development and has established a Foundation for the Transfer of Knowledge, which provides speakers for regional or national conferences.

As a result of deliberations in committees and councils, international associations have incorporated a variety of activities for developing countries, with the following objectives:

- to improve the professional competence of its members,
- to encourage members from developing countries to participate in the process of exchanging scientific information, and
- to diversify their membership.

For improving the professional basis of their members, the associations have produced monographs or guide books on special subjects. Other activities include holding seminars or workshops on engineering or scientific topics and preparing and distributing monographs and training manuals. International professional associations also provide services on national levels for contacts among professionals, one of which is the encouragement of local professional groups. Some international associations have created national committees. In some countries national committees are set up by the national ministry in charge of water resources, a process that helps ensure continued interest in the committee by the top officials of a ministry.

KNOWLEDGE AND TECHNOLOGY TRANSFER

The cumulative experience of engineers and scientists from all over the world forms the basis on which water resources development rests. The dissemination of this knowledge to colleagues in all parts of the world challenges the professional world community. It has been and is one of the most important tasks of the many universities and international associations in the water field.

For this transfer of information and technology, the scientists or the engineers from a developed region should make every effort to concentrate on the needs of their colleagues in the developing regions by becoming true partners in the efforts. Much can be learned by all those participating in such cooperation.

For sustainable water resources development of a region, professionals are needed who have excellent knowledge of both the social and the natural local conditions. In fact, most hydrological or water resources models that have been successfully applied were developed in response to a

local need, and incorporate the most important features of the local situation. The *laboratory* in which the methods have been tested is the region in which the hydrologist or engineer or water scientist worked. The most competent expert is most likely to be a person who knows his or her country well, who knows its people and their needs, and who has a professional toolbox with the applicable fundamental methods. Hence, international associations should increase participation in the work of scientific and technical committees of engineers from developing countries to promote exchange of ideas and experiences, and to obtain from them local information and special knowledge.

SOME THINGS TO REMEMBER

- Resettlement can involve involuntary moves from ancestral homes and traditional living conditions to areas that are unfamiliar – a cost difficult to measure and rarely fully compensated for by those benefiting from the increased irrigation, hydropower, flood control, and other project purposes.

- To be sustainable, a project must perform reliably and the transition to new technologies and management practices must proceed in an orderly and equitable manner. Continuity and confidence in the new systems are prerequisites for sustainability, as are a proper respect for operation rules and for maintenance of the physical infrastructure.

- In both developed and developing regions, non-governmental organizations concerned with environmental and social issues can play an important part in helping to provide frameworks for conflict resolution. They speak for at least some of the people, and they can help provide a balance between governments and the citizens affected by proposed or even existing development projects.

- Outside assistance can affect attitudes towards the use of multiple smaller more sustainable projects as opposed to much larger and more expensive engineering works, which unless well understood and accepted may not be used effectively by the people for whom they were presumably constructed.

- Although water-related problems are usually local or regional, where rivers or lakes cross international boundaries, or where seas separate countries, the problems of water consumption and pollution become international problems, and as such they require joint approaches toward finding acceptable and effective solutions.

- A necessary condition is that appropriate institutions must exist to carry out planning, operation and maintenance of water resources systems. To be sustainable, a system must satisfy local administrative requirements. Its managers must be trained to manage the system taking into account the customs and legal infrastructure of the region.
- Each nation must inevitably come to depend primarily on its own professionals to provide the know-how and experience required for water resources development.
- A key to the sustainable development of water resources is the existence of sufficiently well-trained personnel in all of the fields that are involved in the development process. In regions or localities where such a capacity does not exist, it must be developed.
- The most competent expert is most likely to be a person who knows his or her country well, who knows its people and their needs, and who has a professional toolbox with the applicable fundamental methods.

12 Conclusion

Sustainability issues are not new issues, nor is sustainability a new concept. Yet the current interest in sustainable development clearly comes from a realization that some of the activities that we who inhabit this earth today perform could be causing irreversible damage that may adversely affect not only our own lives but also the lives of those who follow us.

In many situations the overall goals of achieving sustainable development, conserving environmental and natural resources and alleviating poverty and economic injustice, are compatible and mutually reinforcing. However, there will always be conflicting views on how these overall goals can be met, and tradeoffs will have to be made among the conflicting views and objectives. The challenge for political leaders and professional resource managers is to make the best of situations where these complementarities are real, while remaining aware that there are very real situations that will require difficult decisions and choices if sustainable development is to be achieved.

It is clear that there are many unanswered questions related to the sustainable development and management of any renewable or non-renewable water resource system. No manager of water resources has the luxury of waiting until all these questions are answered. But those involved in managing the resources can still work toward increasingly sustainable levels of development and management. This includes learning how to get more from our resources and how to produce less waste that degrades these resources and systems. New ideas and new technology will have to be developed to achieve more economically efficient and effective recycling and the use of recycled materials. Management approaches that are more non-structural and compatible with the environmental and ecological life support systems must be identified. Better ways of planning, developing, upgrading, maintaining and paying for the infrastructure that permits effective and efficient resource management and provides needed services must also be defined.

PLANNING FOR SUSTAINABILITY

Water resources planning is usually performed to determine how to remove most effectively and efficiently a problem or reduce an existing or potential gap between the demand and the supply of services that can be provided by the water resources and the related infrastructure. Sustainable water resource systems, as we have defined them, are:

> *water resource systems designed and managed to fully contribute to the objectives of society, now and in the future, while maintaining their ecological, environmental, and hydrological integrity*

They must be planned, designed and managed in such a way that the life support system at all biological levels remains functional and that the water and related land resource is not irreversibly degraded over time. This imposes constraints on every stage of development – from project planning to its final operation and management within its overall social and technical system.

The modern way of approaching planning and management problems of this kind is through systems analysis and synthesis: the identification, analysis and evaluation of the interactions of all the components of water resource systems in space and time. These components include the set of structural and non-structural components integrated into a network or system involving relationships between humans and their institutions, nature and technology. For such systems to be sustainable, they must interact smoothly with other subsystems of society, and they must adapt to changes and uncertainties in supplies and demands. Multiple alternatives should be defined and evaluated with respect to overall system performance objectives – in short, the systems approach to sustainable water resources planning.

Today's managers of water resources systems have to consider a large number of often conflicting demands on the available water, and they have to develop and operate their systems under numerous social and legal as well as physical constraints. The various different interests require a decision process involving multiple objectives, multiple decision-makers, multiple users and multiple constituencies and stakeholders. Furthermore, the planning processes for water resource system developments must be matched with planning objectives and constraints for transportation, energy production and other sectors of society and its economy.

SYSTEM MONITORING AND EVALUATION

When a project or development plan becomes operational, an important step often ignored or under-funded is the monitoring and analysis of performance aimed at determining the extent to which the objectives of the project or development plan are being achieved. Such monitoring and analyses may, for example, ascertain how much water from an irrigation project is actually being used, and whether or not any unanticipated impacts on project users or on the downstream groundwater regime have occurred. Performance analyses are indispensable for assessing the effectiveness of any project or development plan. They permit corrective actions to be taken to reduce potentially adverse impacts, and may help reduce the risks of future mistakes.

There is always a natural reluctance to carry out *ex-post facto* evaluations. They require time and painstaking collection of data that yield information only after many years. They are also often done by people who were not involved in the planning phases and who therefore may not know the exact planning objectives. In addition, the operating rules will usually change somewhat to accommodate better the actual or perceived situations. 'Final' arguments against such costly steps are that one does not know exactly what to evaluate, and issues of secondary concern may bias the evaluation procedure. However, even though one cannot expect a system to perform exactly as planned, for it to become sustainable it must be adaptive so as to best suit changing conditions. Monitoring and performance evaluations provide a basis for such adaptation.

ECONOMIC AND FINANCIAL ASPECTS

Inefficient operation and poor maintenance of systems present serious problems in many regions, calling into question the long-term sustainability of the systems that have been created. The problems are often rooted in the weakness of public and private institutions, especially in terms of paying for operation and maintenance through water-user fees or income. Cost recovery is absolutely essential if one is to avoid system deterioration and a decline in levels of service, and consequent wastage of investments.

In addition to cost recovery, economic pricing policies can play an important role in facilitating efficient resource use over time as demands for the resource change. For example, reducing water subsidies that has resulted in under-priced water for irrigation may bring in increased benefits by the allocation of that water to other more valuable uses. Realistic pricing can also provide an increased opportunity to accumulate funds for a water resources infrastructure needed in the future.

Shortage of funds may not only affect project implementation, but also the training and education of the management and operations personnel. Financial support may already be needed on a large scale to promote the transfer of knowledge and appropriate technology in many developing countries.

MULTI-DISCIPLINARY INPUTS

Economists are trained to identify pricing incentives that can influence the way the development, management and use of water resources takes place. Engineers are trained to design, construct, operate and maintain facilities required to implement various water resource development and management plans or policies. Both disciplines are critical for sustainable system planning and operation. Yet economists and engineers are not the only professionals that need to contribute toward achieving a greater level of sustainable development. Individuals skilled in anthropology, in agriculture, in ecology, in government, in law, in planning and in sociology as well as in economics and engineering must come together to understand one another and work together. We must also work with the public, to obtain the necessary collective insight and knowledge.

Finally, it is worth reminding ourselves that those of us who take responsibility for the planning and management of resources must assume leadership in the effort of moving everyone toward a process of environmental resource development and use that becomes increasingly sustainable.

WE THE ENTRUSTED

Sustainable water management involves people – people who plan, organize, coordinate, direct, control and supervise the exploration, development and use of the water and related land resources. We who are here today must learn how to do it in ways that will not unduly limit the availability of these resources for future generations. It is imperative that we water and related land resource managers have the appropriate knowledge, know-how, equipment, authority and responsibility to do the job.

As noted in the summary section of a US National Research Council's Report (1991b), entitled *Toward Sustainability in Agriculture*,

> One of the challenges we face in developing strategies that are truly sustainable is maintaining the resource base – the soil and water that makes life as we know it possible. But the pressures on these resources are extraordinary: five billion people now inhabit the earth, with an additional one billion expected each decade well into the next century. The specter of possible changes in climate adds another level of uncertainty. It is time to ask how we can move "toward sustainability", toward a strategy of natural resource management that supports current populations while leaving future generations an equitable share of the earth's great wealth.

Added to the management complexity is the fact that there are substantial gaps in our basic understanding of the physical, chemical and biological 'ecology' of these (soil and water) systems and of the complex social interactions inherent in managing these resources. An important message from this NRC report (and from United Nations, 1990) is the need for a more holistic approach to the management of our natural resources if we are to achieve increased levels of sustainable development.

Sustainable development and management requires competent people willing to work together to achieve that goal. Covey (1989) pays special attention to what he calls personal trustworthiness and its relationship to effectiveness, and to organizational alignment, managerial empowerment and interpersonal communication. Using the terms 'programs' as a metaphor for water resource systems and 'programmers' for water resources systems managers, i.e., people, we can appreciate it when he writes:

> If you want to improve the program, work first on the programmer; people produce the strategy, structure, systems, and styles of the organization. These (institutions and their functions) are the arms and hands of the minds and hearts of people. The key to creating a total quality organization is first to create a total quality person . . .

In striving for sustainable development in water and related land resources, the effectiveness of any mechanism devised to realize that goal depends ultimately on the quality of the individuals entrusted with pursuing it. Engineers, economists, ecologists, planners and other professionals must be involved, but they can be only part of that involvement. Professionals must work within the social infrastructure of a community or region. Failure to do so will result increasingly in challenges to proposed projects after they have been designed, or even after they are under construction or have been built. Successful collaboration with an informed and involved public can lead to more socially compatible uses of resources and to more creative, appropriate, and hence sustainable, uses of technology for addressing a community's or region's water resource problems or needs.

References

Akowumi, F.A., 1994. Some humanistic perspectives on sustainable water resources development in an African nation. *Proceedings, Conference on Water Management in a Changing World*, Karlsruhe, Germany, June 28–30, 21–32.

Alaerts, G.J., Blair, T.L., & F.J.A. Hartvelt (eds), 1991. *A Strategy for Water Sector Capacity Building*, IHE Report Series 24, IHE, Delft, NL.

Albertson, M.L., 1995. Appropriate technology for sustainable development, presented at *ASCE Division of Water Resources Planning and Management National Conference*, Cambridge, MA.

Andreu, J., Capilla, J., and E. Sanchis, 1996, AQUATOOL, a generalized decision-support system for water resources planning and operational management, *Journal of Hydrology*, **177** (3–4), April, 269–291.

Ang, A.H.,& W.H. Tang, 1984. *Probability Concepts in Engineering Planning and Design*, Vol. 2, John Wiley, New York.

Baan, J.A., 1994. Evaluation of water resources projects on sustainable development. *Proceedings, Conference on Water Management in a Changing World*, Karlsruhe, Germany, June 28–30, IV, 63–72.

Bardossy, A., & H.J. Caspary, 1990. Detection of climate change in Europe by analysing European atmospheric circulation pattern from 1881 to 1989. *Theoretical and Applied Climatology*, **42**, 155–167.

Barendregt, A., & J.W. Nieuwenhuis, 1993. ICHORS, Hydro-ecological relations by multi-dimensional modeling of observations. *Proceedings of a Conference on the Use of Hydro-ecological Models in the Netherlands*, CHO-TNO, No. 47, Delft, NL, pp. 11–30.

Behrens, J.S., & Y. Reinink, 1994. Application of the PCRSS Reservoir Simulation Model to the Salt River Project, *Proceedings of the 21st Annual Conference of the Water Resources Planning and Management Division*, American Society of Civil Engineers, Denver, CO, May, pp. 295–98.

Biswas, A.K., Jellali, M., & G. Stout (eds), 1993. *Water for Sustainable Development in the Twenty-first Century*. Oxford University Press, Delhi, India.

Brooks, H., 1992. Sustainability and technology. *Science and Sustainability*, Chapter 1, IIASA, Laxenburg, Austria.

Brown, G.M., Jr., 1991. Can the Sustainable Development Criterion Adequately Rank alternative Equilibria? Mimeo, Department of Economics, University of Washington, Seattle, WA.

Brown, R., 1995. The Tchelo Djegou water-wheel. *Proceedings, ASCE Division of Water Resources Planning and Management*, National Conference, Cambridge, MA, May, pp. 516–19.

Bruce, J.P., 1992. *Meteorology and Hydrology for Sustainable Development*. World Meteorological Organization No. 769, Secretariat of the WMO, Geneva, Switzerland.

Casti, J.L., 1989. *Alternate Realities: Mathematical Models of Nature and Man*. John Wiley, New York.

Casti, J.L., 1992. *Reality Rules: II, Picturing the World in Mathematics – The Frontier*. John Wiley, New York.

Charles, Prince of Wales, 1992. Keynote address to the World Commission on Environment and Development, London, April 22.

Covey, S.R., 1989. *The Seven Habits of Highly Effective People*. Simon & Schuster, New York.

da Cunha, L.V., 1989. Sustainable development of water resources. *Proceedings, International Symposium on Integrated Approaches to Water Pollution Problems*, Lisbon, pp. 3–30.

da Cunha, L.V., 1994. The Aral Sea Crisis. Mimeo. NATO ARW, Skopelos, Greece, May

Delft Hydraulics Lab., 1994. *Framework of Analysis for River Basin Management*. River Basin Management Group, Delft, NL, December.

Diersch, H.-J.G., 1993. Computational aspects in developing an interactive 3D groundwater transport simulator using FEM and GIS. *Paper presented at the International Conference on Groundwater Quality Management (GQM 93)*, Tallinn, Estonia, September.

Diersch, H.-J.G., & S.O. Kaden, 1994. *The Simulation System FEFLOW, User Manual*. WASY Ltd., Berlin.

Dimentman, H.J., & F.D. Bromley, 1992. *Lake Hula: Reconstruction of the Fauna and Hydrobiology of a Lost Lake*. Israeli Academy of Sciences and Humanities.

Dornier, 1993. Ökologischer Sanierungs und Entwicklungsplan Niederlausitz. FE-Vorhaben 10104057/02 des Umweltbundesamtes, Berlin (Ecological recreation and development plan Lower Lusatia research report).

Dziegielewski, B., Garbharran, H.P., & J.F. Langowski, 1993. *Lessons Learned from the California Drought (1987-1992)*. IWR Report 93-NDS-5, Institute for Water Resources, U.S. Army Corps of Engineers, Alexandria, Virginia.

Dziegielewski, B., Lynne, G.D., Wilhite, D.W. & D.P. Sheer, 1991. *National Study of Water Management During Drought: A Research Assessment*. IWR Report 91-NDS-3, Institute for Water Resources, U.S. Army Corps of Engineers, Alexandria, Virginia.

ECLAC, 1989. Guidelines for the analysis of water resource management procedures in Latin America and the Caribbean (Based on Peruvian experiences). United Nations Economic Commission for Latin America and the Caribbean, LC/G.1522, 26 April 1989.

Engelman, R., & P. LeRoy, 1993. *Sustaining Water: Population and Future of Renewable Water Supplies*. Population and Environment Program, Population Action International, Washington, DC.

European Bank for Reconstruction and Development, 1993. *The Danube: A Heritage under Pressure*. Environmental Programme for the Danube River Basin, Draft Summary Report, Equipe Cousteau. Paris, France. April.

Falkenmark, M., 1988. Sustainable development as seen from a water perspective. In *Perspectives of Sustainable Development*, Stockholm Studies in Natural Resources Management, No. 1, Stockholm, pp. 71–84.

Fedra, K., Weigkricht E., & L. Winkelbawer, 1993. *Decision Support and Information Systems for Regional Development Planning*. IIASA RR-93-13, Laxenburg, Austria.

Fedra, K., & D.G. Jamieson, 1996. WaterWare decision support system for river-basin planning. 2. Planning capability, *Journal of Hydrology*, **177** (3–4), April, 177–198

Fitz, H.C., DeBellevue, E.B., Costanza, R., Boumans, R., Maxwell, T., Wainger, L., & F.H. Sklar, 1996. Development of a general ecosystem model for a range of scales and ecosystems, Ecological Modeling, **88** (1–3), July, 263.

Flyvbjerg, B., 1996. Practical philosophy for sustainable development: the phronetic imperative, in M. Rolen (ed.) *Culture, Perceptions, and Environmental Problems: Interscientific Communication on Environmental Issues*, Swedish Council for Planning and Coordination of Research, Box 7101, S-10387, Stockholm, Sweden, pp. 89–109.

Galloway, G.E., Jr., 1994. Floodplain management: a present and 21st century imperative, in sharing the challenge: the next steps. *Water Resources Update*, **97**, 4–8.

German Federal Ministry for the Environment, 1994. *Environmental Policy: German Strategy for Sustainable Development*. Bonn, Germany.

Gleick, P.H., Loh, P., Gomez, S., & J. Morrison, 1995. *California Water 2020: A Sustainable Vision*. Pacific Institute for Studies in Development, Environment, and Security, Oakland, CA.

Golden, H.G., & H.F. Lins, 1990. Drought in the Southwestern United States, 1985–86. *Drought Water Management*, N.S. Grigg and E.C. Vlachos (eds), Report for the Natural and Man-Made Hazards Mitigation Program of the National Science Foundation, Colorado State University, Fort Collins, CO.

Goodland, R., Daly, H., & S. El Serafy, 1991. Environmentally Sustainable Economic Development: Building on Brundtland. World Bank Environmental Working Paper No. 46, World Bank, Washington, DC.

Haimes, Y.Y., 1992. Sustainable development: a holistic approach to natural resources management. *IEEE Transactions on Systems, Man, and Cybernetics*, **SMC3**, 413–417.

Hashimoto, T., Loucks, D.P. & J.R. Stedinger, 1982, Robustness of water resources systems. *Water Resources Research*, **18**(1), February, 21–6.

HELCOM, 1993. The Baltic Sea Joint Comprehensive Environmental Action Programme, Balt. Sea Environ. Proc. No. 48. Helsinki Commission, Helsinki Government Printing Center. Helsinki, Finland.

Holling, C.S. (ed.), 1978. *Adaptive Environmental Assessment and Management*. John Wiley & Sons, New York.

Hooghart, J.C., & C.W.S. Posthumus, 1993. The use of hydro-ecological models in the Netherlands. *Proceedings, Technical Meeting 51, TNO Committee on Hydrological Research*, Ede, The Netherlands.

Horowitz, M.M., 1994. The management of an African river basin: alternative scenarios for environmentally sustainable economic development and poverty alleviation. *Proceedings, Conference on Water Management in a Changing World*, Karlsruhe, Germany, June 28–30. pp IV-73–82.

Hufschmidt, M.M., & K.G. Tejwani, 1993. *Integrated Water Resources Management: Meeting the Sustainability Challenge*. UNESCO IHP Humid Tropics Programme Series No. 5, UNESCO, Paris, France.

Institute for Water Resources, 1991. *The National Study of Water Management During Drought: Report on the First Year of Study*. IWR Report 91-NDS-1, U.S. Army Corps of Engineers, Alexandria, Virginia.

Institute for Water Resources, 1994. *Managing Water for Drought*. IWR Report 94-NDS-8, U.S. Army Corps of Engineers, Alexandria, Virginia.

Institution of Engineers, Australia, 1989. *Policy on Sustainable Development*. Barton, ACT, Australia, July.

International Commission for the Protection of the Rhine, 1994. *Ecological Master Plan for the Rhine: Salmon 2000*. ICPR, Koblenz, Germany.

Jamieson, D.G. & K. Fedra, 1996a. WaterWare decision support system for river-basin planning. 1. Conceptual design. *Journal of Hydrology*, **177** (3–4), April, 163–175.

Jamieson, D.G. & K. Fedra, 1996b. WaterWare decision support system for river-basin planning. 3. Example applications. *Journal of Hydrology*, **177** (3–4), April, 199–211.

Jordaan, J., Plate, E.J., Prins, E., & J. Veltrop, 1993. *Water in Our Common Future: A Research Agenda for Sustainable Development of Water Resources*. Committee on Water Research (COWAR), IHP, UNESCO, Paris, France.

Kaden, S.O., Hänel, U., & Karl-Heiz Seidel, 1976. Digitale Berechnung der Entwässerungswirkung von Tagebaufeld und Randgalerien. (Digital computation of mine drainage with well galleries), Neue Bergbautechnik, Leipzig, 6, June 1976, S. 418–423.

Kaden, S.O., & M. Schramm, 1993. Ökologischer Sanierungs und Entwicklungsplan Niederlausitz, Teilbericht zu wasserwirtschaftlichen Untersuchungen. (Ecological recreation and development plan Lower Lusatia; sub-report on water management studies), FE-Vorhaben 10104057/02 des Umweltbundesamtes, Berlin.

Kaden, S.O., 1994. Groundwater-related decision support systems – tools for sustainable development? *Proceedings, Conference on Water Management in a Changing World*, Karlsruhe, Germany, June 28–30. pp III–15 – III-39.

Kaplan, S., & B.J. Garrick, 1981. On the quantitative definition of risk. *Risk Analysis*, **1**(1), 11–27.

Kerr, R.A., 1992. A Successful forecast of an El Niño winter. *Science*, **255**, 402.

Kunreuther, H.C., & G.F. White, 1995. The role of the National Flood Insurance Program in reducing losses and promoting wide use of floodplains. In Coping with the Flood. The Next Phase, *Water Resources Update*, **95**, 31–35.

L'Vovich, M.I., 1977. World water resources present and future. *Ambio* **6**(1), 13–21.

Larsen, P., 1992. Restoration of river corridors. *Rivers Handbook*, Vol. 2, Chapter 5, by P. Calow and G.E. Petts (eds), Blackwell Scientific Publications, Oxford, pp. 419–38.

Loucks, D.P., & J.R. da Costa, (eds), 1991. *Decision Support Systems: Water Resources Planning*, Springer Verlag, Berlin, Germany.

Loucks, D.P., French, P.N., & M..R. Taylor, 1995. Interactive River-Aquifer Simulation: Program Description and Operation, Cornell University, Ithaca, NY.

Lowrance, W.W., 1976. *Of Acceptable Risk*. William Kaufman, Inc., Los Altos, CA.

Matheson, S., 1997. Distributive Fairness Criteria for Sustainable Project Selection, M.Sc.Thesis, Department of Civil and Geological Engineering, University of Manitoba, Winnipeg, Canada.

Marco, J.B., & N. Cayuela, 1994. Urban flooding: the flood planned city. In *Coping with Floods*, G. Rossi, N. Hanmancioglu, and V. Yevjevich (eds), Kluwer Publishers, The Netherlands.

Maxwell, T., & R. Costanza, 1994. Spatial ecosystem modeling in a distributed computational environment. In *Toward Sustainable Development: Concepts, Methods and Policy*, J.C.J.M. van den Bergh and J. van der Straaten (eds). Island Press, Washington, DC.

McDonald, A.T., & D. Kay, 1988. *Water Resources Issues and Strategies*. Longman Scientific & Technical, Essex, UK.

McMahon, G.F., & J.R. Mrozek, 1996. Economics, Entropy and sustainability. *Hydrologic Sciences Journal*. **42**(4), August, 501–38.

Meadows, D.H., Meadows, D.L., Randers, J., & W.W. Behrens, III, 1974. *Limits to Growth*. Report for the Club of Rome's Project on the Predicament of Mankind, 2nd. edition. Universe Books, NY.

Morse, B., & T.R. Berger, 1992. *Sardar Sarovar*. Resources Futures International, Ottawa, Canada.

Munasinghe, M. (ed.), 1993. *Environmental Economics and Natural Resources Management in Developing Countries*. World Bank, Washington, DC.

Munasinghe, M., & E. Lutz, 1991. *Environmental-Economic Evaluation of Projects and Policies for Sustainable Development*. Environmental Working Paper No. 42, World Bank, Washington, DC. January.

Nachtnebel, H.P., 1996. Irreversibility and sustainability in water resource systems, *Third IHP/IAHS Kovacs Colloquium*, UNESCO, Paris, France, September.

Naiman, R.J., Magnuson, J.J., McKnight, D.M., & J.A. Stanford, 1995. *The Freshwater Imperative: A Research Agenda*. Island Press, Washington, DC.

Norgaard, R.B., & R.B. Howarth, 1991. Sustainability and discounting the future. In *Ecological Economics: The Science and Management of Sustainability*, R. Costanza (ed.), Columbia University Press, New York, NY, pp. 88–101.

Okun, D.A., 1991. A water and sanitation strategy for the developing world. *Environment*, **33** (8), October, 16–20.

Orlovski, S., Kaden, S., & P. van Walsum, 1986. *Decision Support Systems for the Analysis of Regional Water Policies*. Final Report, WP-86-33, IIASA, Laxenburg, Austria.

Pearce, D.W., 1993. Valuing the environment: past practice, future prospect. *First Annual International Conference on Environmentally Sustainable Development*, IBRD, Washington, DC, September 30 – October 1.

Pearce, D.W., & J.J. Warford, 1993. *World Without End. Economics, Environment and Sustainable Development*, The Oxford University Press, Oxford, UK.

Pearce, D., Markandya, A., & E.D. Barbier, 1989. *Blueprint for a Green Economy*. Earthscan, London.

Pennekamp, H.A., & J.W. Wesseling, (eds), 1993. *Methodology for Water Resources Planning*. Delft Hydraulics Lab. Delft, NL.

Pethig, R., 1994. *Valuing the Environment: Methodological and Measurement Issues*. Kluwer Academic Publishers, Dordrecht, The Netherlands.

Pezzey, J., 1992. *Sustainable Development Concepts, An Economic Analysis*. World Bank Environment Paper Number 2, World Bank, Washington, DC.

Plate, E.J., 1991. Die Auswirkung von Klimaanderungen auf Sturmfluten. *Hansa-Schiffahrt*, Schiffbau-Hafen, **128**, 1174–1181.

Plate, E.J., 1992. Scientific and Technological Challenges. Keynote Paper No. 10 for the ICWE Secretariat, International Conference on Water and the Environment, *Keynote Papers*, Dublin, Ireland, WMO, Geneva, Switzerland (prepared by COWAR: the Committee for Water Research of the International Council of Scientific Unions (ICSU) and of the Union of International Technical Associations (UITA).

Prendergast, J., 1993. Engineering sustainable development. *Civil Engineering*, **63**(10), October, 39–42.

Reitsma, R.F., 1996. Structure and support of water resources management and decision making, *Journal of Hydrology*, **177** (3–4), April, 253–68.

Rich, J., 1993. Institutional Responses to the 1987–92 California Drought. Chapter 14 of *Drought Assessment, Management, and Planning: Theory and Case Studies*, D.A. Wilhite, ed., Kluwer Academic Publishers, Boston, MA.

Rifkin, J., 1989, *Entropy: Into the Greenhouse World*, Bantam Books, New York, NY.

Rotmans, J., van Asselt, M.B.A., de Bruin, A.J., den Elzen, M.G.J., de Graf, J., Hilderink, H., Hoekstra, A.Y., Janssen, M.A., Koster, H.W., Martens, W.J.M., Niessen, L.W., & H.J.M. de Vries, 1994. *Global Change and Sustainable Development: A Modelling Perspective for the Next Decade*. Global Dynamics and Sustainable Development Programme, National Institute of Public Health and Environmental Protection, RIVM Report No. 461502004, Report Series No. 4, Bilthoven, NL.

Ruth, M., 1993, *Integrating Economics, Ecology and Thermodynamics*, Kluwer Academic Publishers, Dendrich.

Saeijs, H.L.F., van Westen, C.-J., & M.H. Winnubst, 1995. Time for a revival of the Rhine: a quest for a sustainable river basin. *Reservoirs in River Basin Development*, Santbergen and van Westen (eds), Balkema, Rotterdam, NL.

Salem, O.M., 1992. The Great Manmade River Project. *Water Resources Development*, **8**(4), December, 270–8.

Schmid, A.S., 1993. The interdisciplinary significance of water in the ecological renewal of the Emscher Region. In *HYDROPOLIS, International Workshop*, Wageningen, Netherlands, March 29–April 2.

Serageldin, I., Mink, S., Cernea, M., Rees, C., Munasinghe, M., Steer, A., & E. Lutz, 1993. Sustainable Development. A series of articles on sustainable development in *Finance & Development*, **30**(4), Published by the World Bank, Washington, DC, December.

Simonovic, S.P., 1996. Decision support systems for sustainable management of water resources, *Water International*, **21**(4), 223–44.

Sinha, S.K., Kailasanathan, K., & A.K. Vasistha, 1987. Drought management in India: steps toward eliminating famines. In *Planning for Drought: Toward a Reduction of Societal Vulnerability*, D.A. Wilhite and W.E. Easterling (eds), Kluwer Academic Publishers, Boston, MA.

South Florida Water Management District, 1988. Kissimmee River Restoration Project, *Closer Look* newsletter, West Palm Beach, FL.

Spence, J., 1997. A Flood of Troubles, *New York Times Magazine*, January 5, 34–9.

Steuer, R.E., 1986. *Multiple Criteria Optimization: Theory, Computation, and Application*. John Wiley & Sons, New York.

Svedin, U., 1988. The Concept of Sustainability. In *Perspectives of Sustainable Development*, Stockholm Studies in Natural Resources Management, No. 1, Stockholm, pp. 1–18.

Taylor, M.R., & J. Behrens, 1996. NSF Proposal On Object Oriented Programming for Hydrologic and Ecologic Modeling, Resources Planning Assoc., Inc., Ithaca, NY.

Toman, M.A., & O. Crosson, 1991. Economics and "Sustainability", Balancing Trade-offs and Imperatives, ENR91-05, Resources for the Future, mimeo, Washington, DC.

Traore, A., 1992. Community Water; Drinking Water Supply and Sanitation in the Rural Context. Keynote paper (actual presentation), *International Conference on Water and the Environment (ICWE)*, Dublin, Ireland, January 26–31.

United Nations, 1989. *World Population Prospects*. Department of international Economic and Social Affairs, New York, NY.

United Nations, 1990. *An International Action Programme on Water and Sustainable Agricultural Development: A strategy for the implementation of the Mar del Plata Action Plan for the 1990*. Food and Agriculture Organization (FAO), Rome, Italy.

United Nations, 1991. *A Strategy for Water Resources Capacity-building: The Delft Declaration*. UNDP Symposium, Delft, June 3–5.

United Nations, 1992. *Report on Water Quality: Progress in the Implementation of the Mar del Plata Action Plan and a Strategy for the 1990's*. World Meteorological Organization and United Nations Environmental Program (WMO, UNEP).

United Nations Conference on Environment and Development (UNCED), 1992. Rio de Janeiro, Brazil, 3–14 June.

UNESCO, 1991. *The Disappearing Tropical Forests*. IHP Humid Tropics Programme Series No. 1, Paris, May 1991.

UNESCO, 1992. *Water and Health*. IHP Humid Tropics Programme Series No. 3, Paris, France, May.

US National Research Council (NRC), 1991a. *Opportunities in the Hydrological Sciences*. National Academy Press, Washington, DC.

US National Research Council (NRC), 1991b. *Toward Sustainability in Agriculture*, National Academy Press, Washington, DC.

US Water Resources Council, 1973. *Water and Related Land Resources; Establishment of Principles and Standards for Planning*. Washington, DC.

UVPG, 1990. Gesetz über die Umweltverträglichkeit-sprüfung vom 12.2.1990; BGB1. I 2.205.

Valdez, J.B., Marco, J.B., Wurbs, R.A. & A. Mejia, 1995. Water Resources Extremes and Sustainable Development. Mimeo. Civil Engineering, Texas A&M University, College Station, Texas.

van den Bergh, J.C.J.M., & J. van der Straiten (eds), 1994. *Toward Sustainable Development*. Island Press, Washington, DC.

van Dijk, G.M., van Liere, L., Admiraal, W., Bannink, B.A., & J.J. Cappon, 1994. Present state of the water quality of European rivers and implications for management. *The Science of the Total Environment*, **145**, 187–95.

van Dijk, G.M., & E.C.L. Marteijn (eds), 1993. Ecological Rehabilitation of the River Rhine: 1988–1992, Report No. 50, RIZA, Lelystad, NL.

van Dijk, G.M., Marteijn, E.C.L., & A. Schulte-Wulwer-Leidig, 1995. Ecological rehabilitation of the River Rhine: plans, progress and perspectives. *Regulated Rivers: Research and Management*, **11**(3–4), 377–88.

Vlachos, E.C., 1990. Unfolding events of the drought: The socio-economic context. *Drought Water Management*, N.S. Grigg and E.C. Vlachos (eds). Report for the Natural and Man-Made Hazards Mitigation Program of the National Science Foundation, Colorado State University, Fort Collins, CO.

Waterstone, M., 1994. Institutional approaches to water management under conditions of global change, presented at *Conference on Water Management in a Changing World*, Karlsruhe, Germany, June 28–30.

WCED (World Commission on Environment and Development), 1987. *Our Common Future*. ("The Brundtland Report"), Oxford University Press.

Wilhite, D.A., 1993. Enigma of drought, Chapter 1 and Planning for drought: a methodology. Chapter 6, *Drought Assessment, Management, and Planning: Theory and Case Studies*, D.A. Wilhite, (ed.), Kluwer Academic Publishers, Boston, MA, pp. 1–15 and pp. 87–108.

World Bank, 1993. *Water Resources Management*, International Bank for Reconstruction and Development, Washington, DC.

World Bank, 1994. *Making Development Sustainable*. International Bank for Reconstruction and Development, Washington, DC.

Young, M.D., 1992. *Sustainable Investment and Resource Use: Equity, Environmental Integrity and Economic Efficiency*. UNESCO Man and the Biosphere Series Vol. 9, UNESCO and Parthenon Publishing Group, Paris, France.

Index

adaptive planning *x*, 58, 60, 61, 66, 77, 79, 80, 83, 84, 86, 88, 89, 90, 91, 97, 99, 100, 107, 123, 132
alternatives (dominated/non-dominated) 36, 72

benefit–cost analyses 81
biodiversity 3, 5, 9, 10, 29, 56, 58, 108
bottom-up approach 17, 54, 57, 66, 84
Brundtland Commission, *Our Common Future* 6, 7, 10, 26

capacity building *xi*, 5, 22, 25
carrying capacity 29, 33
case study
  Aral Sea (Central Asia) 45–6
  Central Utah Project (USA) 57
  Columbia River (USA) 58–60
  Danube River Basin (Europe) 20, 22, 50–2, 96
  Great Lakes Basin (Canada and USA) 54–5
  High Aswan Dam (Egypt) 58
  Hula Valley Project (Israel) 61–2
  Hydropower in Malaysia 58, 128
  Kissimmee River (USA) 60–1
  Mekong River (Southeast Asia) 62–4
  Milwaukee (USA) 62
  North and Baltic Sea (Europe) 14, 20, 52–3
  Ogallala Aquifer (USA) 17, 46–8
  Olympus Dam (USA) 57
  Orangi Pilot Project of Karachi (Pakistan) 53–4
  Rhine River (Europe) 20, 22, 49–50
  Sahara in Libya (Libya) 48–9
  Senegal River (Senegal, Mauritania, Mali) 55–7
  Tarbella Dam (Pakistan) 57
  Tchelo Djegou (Niger) 57–8
  Volta River (Ghana) 64–5
climate change 10, 21, 23, 24, 33, 94, 112, 114, 127
Codes of Practice 30
common property 10, 71
communication *x*, 84, 85, 86, 87, 100, 101, 133
  Anacostia, Maryland (USA) 85
  Cape Cod, Massachusetts (USA) 85
  Connecticut wetlands (USA) 85
  Cordova, Alaska (USA) 84–5
conjunctive use 18, 25
consensus-based resource management 84

consumptive/non-consumptive uses 6, 67, 68, 69, 119
criteria
  ecological *x*, 34, 41, 81, 82, 92, 97
  economic *x*, 34, 41, 92, 97
  environmental *x*, 34, 41, 81, 92, 97
  social 2, 34, 41, 97

decision support systems (DSSs) *xi*, 4, 99–107
  eastern Germany 101–4
  European Union 100–1
  modules and links 104–6
demand management 4, 15–16, 30, 43, 115, 116, 121
disaster management 21, 109–10
discount rate *x*, 32, 68, 75, 77, 78
discounting 10, 32, 78
drought policies 116–17

ecologically conscious 28
economic efficiency 29, 32–3, 34, 41, 43, 73, 75, 77, 78, 79, 112, 116
ecosystem approach 43, 54, 59, 84, 86, 87, 89
ecosystem integrity 29, 55
environmental degradation 14, 16, 69
environmental impact 6, 7, 8, 9, 14, 22, 43, 58, 64, 68, 69, 70, 71, 81–3, 88, 96, 100, 101, 102, 113, 114, 119
environmental protection 18, 32, 62, 111
equity 9, 11, 28, 29, 31, 46, 122–6,
excess water 19
expected maximum risk 111
expected risk 110, 111
expected value (conditional/unconditional) 36, 111

fairness, intergenerational 123–6
  intragenerational 123–6
  measure 123–6
floods and droughts 14, 20–1, 22, 37, 39, 50, 112–16
flood protection 69, 94, 112,
future uncertainty 108

general circulation models (GCMs) 23
global dimensions 12
global impacts 6, 7, 10, 24
groundwater management 17–18, 101–4
groundwater mining 76–9
groundwater protection 17
guidelines, samples
  design, management and operation *x*, 3–4, 7, 15, 30, 42–3

economics and finance *x*, 3–4, 7, 15, 30, 42, 43
  environment and ecosystems *x*, 3–4, 7, 15, 30, 42, 43
  health and human welfare *x*, 3–4, 7, 15, 30, 42, 44
  institutions and society *x*, 3–4, 7, 15, 30, 42, 43–44
  planning and technology *x*, 3–4, 7, 15, 30, 42, 44

hydrologic hazards and disasters 109–10
hydropower 7, 8, 18–19, 20, 34, 50, 56, 57, 58, 59, 64, 68, 69, 94, 95, 96, 98, 99, 115, 117, 119, 122, 129

information and data management 83–4, 87
institutional analysis 93
institutional arrangements
  Gabcikovo Dam (Hungary and Slovakia) 96
  Rio Grande (USA, Mexico) 94–5
  Yangtze (PR of China) 95–6
institutional constraints 83, 91
institutional levels 92, 93, 94
institutions and change 92–4
interagency coordination 86
interest groups (lobbies) 94
intergenerational aspects 10, 24, 29, 31
internalizing environmental externalities 68–9
International Drinking Water and Sanitation Decade 16

knowledge and technology transfer *xi*, 122, 126–7, 128, 129

life-cycle 8, 29, 61
limits to sustainability
  financial 13–14
  water resources 12–13

mega-cities 17, 127
modeling 98–100, 106–7
  DSS 104–106
  eastern Germany 101–4
  European Union 100–1

natural disasters *ix*, 3, 4, 8, 11, 14, 17, 20, 21, 22, 23, 34, 36, 40, 41, 50, 56, 57, 61, 69, 70, 71–2, 76, 79–80, 81, 95, 96, 98–100, 109, 111, 112, 113–20, 121, 127
natural ecosystems 12, 18, 28–9, 34, 43, 81, 83, 113, 127

navigation 7, 20, 21, 34, 58, 62, 69, 82, 95, 117
non-governmental organizations (NGOs) 54, 128

planning techniques, advances in 98–106
pollution (point/non-point) 16, 18, 19, 52, 95, 104
precautionary principle 24
preservation 3, 26, 27, 29, 30, 61–2, 76, 80, 96, 107
professional societies 128–9
public education 115–16
public participation x, 28, 29, 57, 66, 84–6, 93, 106, 120
    see also stakeholders
public/private partnerships 126–7

real-time planning 4, 99, 107
regret 27, 61
reliability 3, 11, 34–40, 41, 58, 82, 110
remedial measures 5, 18, 25
resilience 3, 11, 29, 33, 34–40, 41, 42
resource allocations 16, 19, 48, 67, 74, 77–9, 111
reversibility 40–1
risk and uncertainty ix, xi, 90, 108–21
risk management 4, 110–12, 115, 116, 121
risk mitigation, drought
    California, USA 118
    Peru 120
    Segura River basin, Spain 118–20
    Southeast USA 117

Western USA 117
risk-based decision-making 110–12
robustness x, xi, 2, 8, 13, 21, 23, 40–1, 61, 64, 99, 103, 111

scales 2–3, 5, 11, 14–15, 25, 26, 30, 60
scorecard 36, 37, 38–9, 40, 41
sensitivity analyses 98, 104, 108, 111, 114
shared vision x, 3, 54, 83, 84–6, 89, 105, 116
social responsibilities 91–2
stakeholders ix, x, 3, 4, 7, 8, 40, 43, 44, 54, 57, 58, 61, 66, 83, 84–6, 87, 89, 98, 105, 106, 107, 110, 116, 132
strategic planning 85, 111
substitution of natural resources 2, 9, 11, 25, 26, 33
survivability 32–3
sustainability and risk 4–5
sustainability and technology 4
sustainability and training 5
sustainability defined (water resources) 30
sustainability index 34, 36–40
sustainability issues
    agricultural and industrial demand 19–20
    capacity building 22–3
    environmental protection 18
    groundwater management and use 17–18
    hydroelectric energy 18–19
    natural and man-made environments 22
    natural disaster 21–2
    reservoir storage and operation 21
    time and space scales 14–15
    transport and flood protection 20–1

water quality and health 16–17
water supply and demand management and population growth 15–16
system monitoring and evaluation 3, 4, 13, 16, 43, 83, 88–9, 90, 132

top-down management 17, 25
tradeoffs ix, x, 1, 4, 5, 10, 11, 24, 25, 27, 31, 33, 34, 36, 55, 64, 71, 72, 73, 75, 76, 79, 80, 86, 91, 96, 111, 115, 121, 131

UN Conference on Environment and Development (UNCED) 6, 7

valuing the environment, artificial markets 70, 71
    contingent valuation method 70, 71
    defensive expenditures 70
    differential method 70, 71
    existence value 67, 68, 71
    implicit or surrogate markets 70, 71, 92
    intrinsic (amenity) values 67, 68, 69
    property value method 70, 71
    replacement costs 2, 12, 23, 70, 71
    shadow project approach 70, 71
    value-of-life approach 70
vulnerability 3, 4, 11, 14, 33, 34–40, 41, 86, 114, 116, 121

weighted criteria indices 33–4
weighted statistical indices 34–40